Organic Geochemistry in Petroleum Exploration

Colin Barker
University of Tulsa

Acknowledgements

The AAPG Continuing Education Course Note Series
is an author – prepared publication of the
AAPG Department of Education

Extra copies of this, and all other books in the AAPG
Education Course Note Series, are available from:

AAPG Bookstore
P.O. Box 979
Tulsa, Oklahoma 74101 – 0979

Published March 1979
Fourth Printing, January 1986
ISBN: 0 – 89181 – 159 – 1

C O N T E N T S

PREFACE

Organic geochemistry is now sufficiently well developed that it can be a useful addition to the range of techniques available to the exploration geologist. The recent growth of service companies has made geochemical data readily available without the need for investing in specialized equipment or experienced personnel.

These course notes (and the associated lectures) are intended to introduce the fundamental ideas of organic geochemistry to exploration geologists. The course has been designed with four objectives. First, to let an explorationist know what organic geochemistry can do for him (or her), and equally important, what it cannot do. Secondly, to provide an understanding of the language of organic geochemistry so that he can read and understand published papers, technical presentations and service company (or in-house) reports, and also be able to talk to geochemists. Thirdly, to provide some insights into general concepts such as minimum time and temperature for petroleum generation, minimum organic matter content for a source rock, maximum temperature for crude oil occurrence and the role of organic matter type in controlling petroleum composition. And finally, to provide information on the techniques available for answering specific questions such as whether two oils are related, whether a particular shale unit could be a source, whether the oil in one field is a bacterially degraded equivalent of oil in an adjacent field, etc. These are ambitious aims for a few hours of lectures! Hopefully these notes will provide additional reference materials where the serious explorationist can look for further discussions and additional geologic examples.

Any compilation of this sort owes a great deal to many people, both through their published work and through informal contacts. I must also acknowledge the audiences who made helpful comments on previous presentations of this material. In addition, I am particularly grateful to Drs. John Comer, Parke Dickey, Tom Ho, John Oehler and Jack Williams who provided helpful comments on the first version of the manuscript. Those errors remaining are, of course, my responsibility.

Permission to reproduce figures was granted by the Canadian Society of Exploration Geologists, Editions Technip, Elsevier, Geological Society of Canada, Oil and Gas Journal and World Oil. This permission is gratefully acknowledged.

TOPIC 1

INTRODUCTION

The job of a petroleum explorationist is to find oil and gas. One measure of his ability is the wildcat success ratio or the number of dry holes drilled for each one that produces commerical petroleum. Wagner and Iglehart (1974) reported that 25,562 new field wildcat wells drilled in the United States between 1969 and 1973 found 572 significant new fields -- those containing more than 1 million barrels of oil or 1 bcf gas (AAPG categories A through E). In other words, only one wildcat in 45 led to significant production in this five-year period.

At present, explorationists rely almost entirely on seismic methods of geophysics to locate subsurface features and define traps. Although "bright-spot" geophysical procedures offer hope of identifying the presence of gas in some types of reservoirs, there is no generally applicable method for deciding whether any particular structure will contain oil, or be empty.

Let us assume for a moment that geophysics will be developed to the point where every subsurface feature can be defined unambiguously, including lithologic changes which control stratigraphic traps. What will be the chance of finding oil if a structure is drilled? Statistically it will be the same as the ratio of the number of traps containing oil to the number that do not. Because many wildcat wells test structures, the wildcat success ratio gives a first approximation to the ratio of productive to dry structures of 1 to 45. However, some wildcat wells fail to penetrate structures. If we assume that this is the case for about one-third of the wells, then the ratio of productive to dry

1

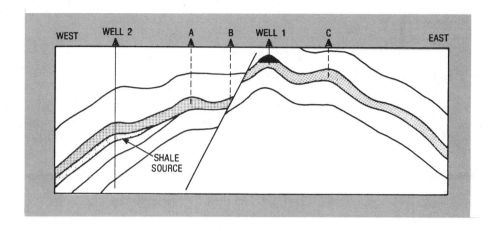

Figure 1.1:
Hypothetical section across an area in the early
stages of development. (Barker, 1975).

structures becomes roughly 1 to 30. This is a very significant number

because it represents the best odds that can ever be achieved for

finding oil in a given structure even after geophysics has been developed

to the ultimate degree of refinement. The actual ratio may be very

different to the one quoted here but the important point is that seismic

methods alone are inherently limited. Obviously, the only way of improving

odds for success is to use additional exploration techniques. Organic

geochemistry is one of these.

The application of organic geochemistry in exploration can be

illustrated by an example. Figure 1 shows an east-west section through

a hypothetical area in the early stages of exploration. Well 1 was

successful in finding oil, but Well 2, some distance to the west, was

dry. If geochemical correlation techniques (based on lab analyses)

showed the shale to be the source rock for the crude oil, then this

immediately indicates the migration path. Any structures along this

path will be highly prospective. The fact that oil found in Well 1 came

from shale penetrated by the dry hole shows that the fault is (or was) not acting as a seal. And from this it follows that the structure under B is not efficiently sealed and is not as good a prospect as the anticline under A. In this particular case, information from the dry hole helps to establish the migration path and indicates the most likely structure for subsequent drilling. Conversely, if the shale could not be correlated to the oil, it would indicate that the fault was probably a seal. In this case both A and B would be equally attractive drill sites. If the crude did not come from the shale to the west, where did it come from? Since the fault appears to be a seal for hydrocarbons moving from the west, Well A's oil probably originated in deeper shales to the east, in which case the structure under C becomes a prime exploration target.

In basins where source rock and crude cannot be correlated, there must be another source rock for the oil, and other oils may have been generated from the source rock. Such a basin would be more attractive than one which has only a single source. Similarly, the presence of two unrelated crude oils shows that there must be at least two source rocks in the area. Any area with multiple sources is attractive (Barker, 1975).

It must be stressed that organic geochemical data should not be used alone but must be integrated with information from other sources. In the example given above a thorough understanding of the geology, and particularly the time of faulting and folding, is needed. In most basins the addition of geochemical data to that obtained by the more conventional methods of geology and geophysics will help improve the chances of success.

In order to apply geochemistry in exploration as outlined in the above example we must have methods for recognizing source rocks and correlating crude oils to each other and to their source rocks. Identification of source rocks requires an understanding of the processes by which they generate crude oils and the way in which the oil migrates. Correlation involves an understanding of the migration mechanism and the factors which can operate to change the composition of reservoired crude oils. Basic to these considerations is a knowledge of the terminology used and an understanding of the environments in which organic matter-rich rocks form. Therefore, the course starts with a brief review of terminology and an outline of organic matter in sedimentary environments and continues by considering the processes which operate to generate petroleum. This is followed by a discussion of the movement of petroleum from the source rock to the reservoir and the nature of the changes which may subsequently occur there. This information provides the background required for a discussion of the techniques for recognizing source rocks and for using geochemical correlation in exploration.

REFERENCES

Barker, C., 1975. Oil source rock correlation aids drilling site selection: World Oil, Oct. 1975, p. 121-126, 213.

Wagner, F. J. and Iglehart, C. F., 1974. North American drilling activity in 1973: Bull. AAPG, v. 58, p. 1475-1505.

TOPIC 2

REVIEW OF FUNDAMENTALS

INTRODUCTION

The terminology of organic geochemistry is, in a very real sense,
the language of organic chemistry. The following outline provides a
brief review of the basic terminology necessary for understanding the
application of organic geochemistry in exploration. The treatment is
not meant to be comprehensive. In particular, it should be noted that
the system used for naming organic compounds is a simplified and abridged
form of several more comprehensive ones.*

The simplest organic compound is methane in which one carbon atom
is linked tetrahedrally to four hydrogen atoms (Figure 2.1). The link-
ages, or bonds, are fixed in direction and length and are covalent in
character. Even in more complex organic compounds carbon always forms
four bonds and these commonly link the carbon to hydrogen, oxygen,
sulfur or to other carbon atoms. This ability to form carbon-carbon
bonds results in large groups of linked carbon atoms and leads to the
enormous number and variety of organic compounds. Representing these
structures presents problems because even the simplest hydrocarbon,
methane, is three-dimensional and must be shown by a perspective drawing
or be drawn in two dimensions according to some convention. Many different methods are in use and some of the more common ones are illustrated

_____

*The interested reader is referred to Banks (1976) or the Handbook of
Chemistry and Physics which give detailed rules for one of the inter-
nationally accepted systems together with well-known synonyms (Weast,
1970).

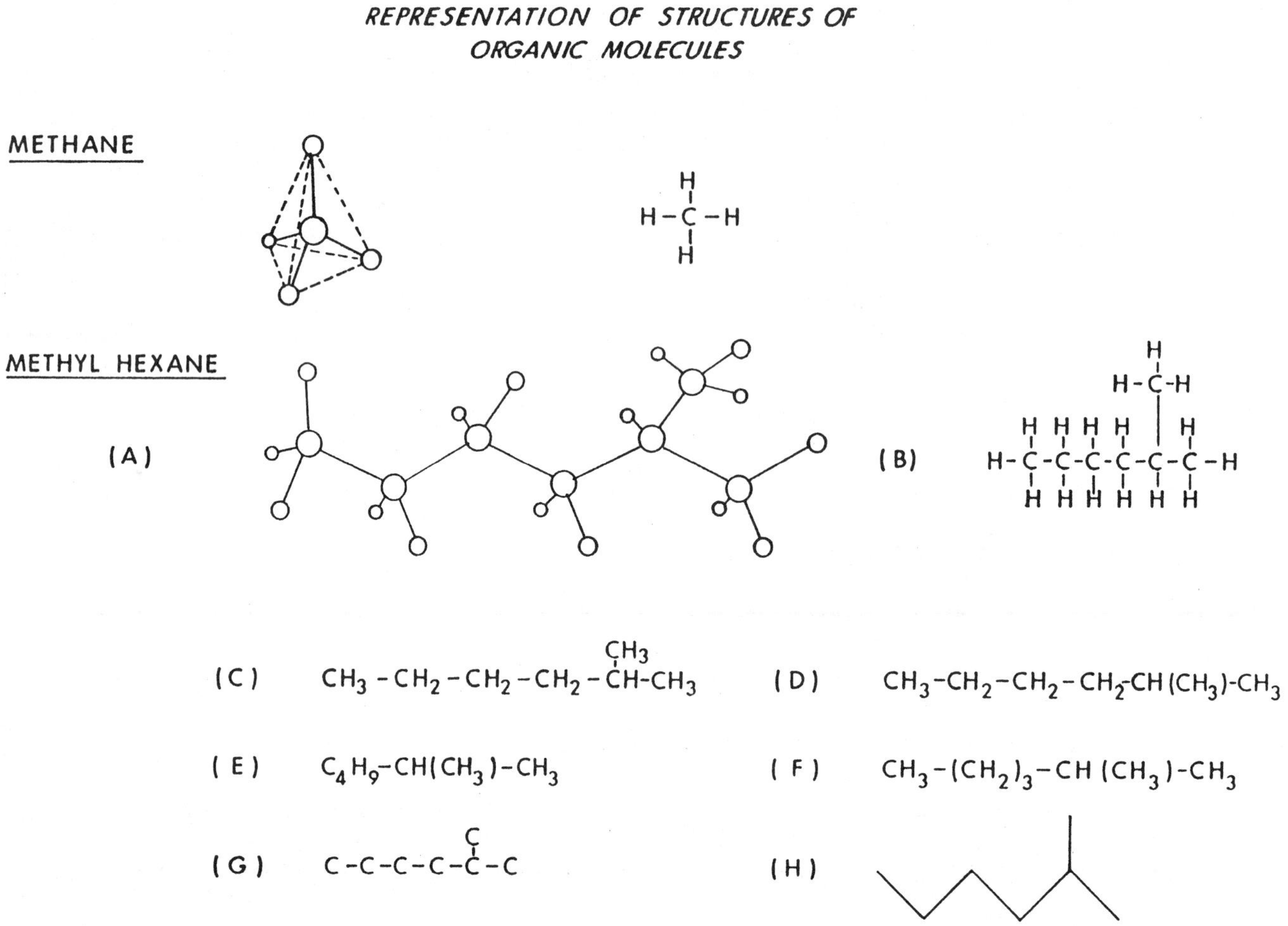

REPRESENTATION OF STRUCTURES OF ORGANIC MOLECULES

METHANE

METHYL HEXANE

(A)

(B)

(C) $CH_3 - CH_2 - CH_2 - CH_2 - \overset{CH_3}{\underset{|}{CH}} - CH_3$

(D) $CH_3 - CH_2 - CH_2 - CH_2 - CH(CH_3) - CH_3$

(E) $C_4H_9 - CH(CH_3) - CH_3$

(F) $CH_3 - (CH_2)_3 - CH(CH_3) - CH_3$

(G) $C - C - C - C - \overset{C}{\underset{|}{C}} - C$

(H)

Figure 2.1:
Some conventions for representing structures
of organic molecules. A through H are all
representations of 2-methyl hexane.

in Figure 2.1 using the branched 7-carbon compound ("2-methylhexane") as

an example.

HYDROCARBONS

Hydrocarbon molecules contain only carbon and hydrogen atoms. The

carbon atoms may be in straight chains, branched chains, rings, or in

combinations of these.

Straight Chains

Hydrocarbons with chain structures are known as paraffins, or alkanes. If the chain is unbranched they are called "normal" and written "n-paraffin" or "n-alkane." The number of carbon atoms in a compound can be indicated by a subscript in the form "C_6", etc. Thus a normal paraffin with fifteen carbon atoms would be abbreviated to "nC_{15}", and the group of compounds with four to seven carbon atoms given as "C_4-C_7". The three simplest normal-paraffins are methane, ethane and propane:

$$
\begin{array}{cccc}
\text{H} & \text{H \quad H} & \text{H \quad H \quad H} & \text{H} \\
| & |\quad| & |\quad|\quad| & | \\
\text{H}-\text{C}-\text{H} & \text{H}-\text{C}-\text{C}-\text{H} & \text{H}-\text{C}-\text{C}-\text{C}-\text{H} & \text{H}-\text{C}-\text{H} \\
| & |\quad| & |\quad|\quad| & | \\
\text{H} & \text{H \quad H} & \text{H \quad H \quad H} & \text{H}
\end{array}
$$

methane   ethane   propane C_nH_{2n+2}

The chain can be lenthened indefinitely by inserting $-CH_2-$ units, giving a series of compounds which all share the general formula C_nH_{2n+2}. As each $-CH_2-$ unit is added, the molecules increase in molecular weight and the melting and boiling points increase (Table 2.1). A group of compounds related in this way is called an "homologous series."

Branched Chains

Branching may occur at one or more of the carbon atoms in a chain. If a branch with a single carbon atom is located on the second carbon in a chain it is called an "isoalkane" or "isoparaffin". In an alternative nomenclature the carbon with the branch is identified by assigning it a number starting from 1 at the end of the chain. Thus an "isoparaffin" would be a "2-methyl paraffin". The seven-carbon hydrocarbon with a branch on the second carbon atom from the end of the six-membered chain would be called "2-methyl hexane" (Figure 2.1).

7

COMMON NAMES & PHYSICAL CONSTANTS OF NORMAL ALKANES

NAME	FORMULA	M.p., ^{0}C	B.p., ^{0}C	STATE AT 1 ATMOS. & 25^{0}C
METHANE	CH_4	-183	-162	GAS
ETHANE	C_2H_6	-172	- 89	
PROPANE	C_3H_8	-187	- 42	
BUTANE	C_4H_{10}	-135	- 0.5	
PENTANE	C_5H_{12}	-130	36	
HEXANE	C_6H_{14}	- 94	69	
HEPTANE	C_7H_{16}	- 91	98	
OCTANE	C_8H_{18}	- 57	126	
NONANE	C_9H_{20}	- 54	151	
DECANE	$C_{10}H_{22}$	- 30	174	LIQUID
UNDECANE	$C_{11}H_{24}$	- 26	196	
DODECANE	$C_{12}H_{26}$	- 10	216	
TRIDECANE	$C_{13}H_{28}$	- 6	234	
TETRADECANE	$C_{14}H_{30}$	6	251	
PENTADECANE	$C_{15}H_{32}$	10	268	
HEXADECANE	$C_{16}H_{34}$	18	280	
HEPTADECANE	$C_{17}H_{36}$	22	303	
OCTADECANE	$C_{18}H_{38}$	28	303	
NONADECANE	$C_{19}H_{40}$	32	330	SOLID
EICOSANE	$C_{20}H_{42}$	36	---	

Table 2.1
Common Names and Physical Constants
of Normal Alkanes

Branched alkanes have the same general formula, C_nH_{2n+2}, as the normal alkanes so that a formula of C_5H_{12}, for example, can correspond to more than one structure. The different forms are called "structural isomers" and for C_5 compounds the possibilities are:

8

```
                   C                            C
                   |                            |
C-C-C-C-C        C-C-C-C                      C-C-C
                                               |
                                               C
```

normal pentane   isopentane   neopentane

(n-pentane)   (2-methylbutane)   (2,2-dimethylpropane)

The number of possible isomers increases very rapidly with the number of

carbon atoms in the compound.   For $C_{30}H_{62}$ there are 4,111,846,783 possible

structures!

"Optical isomers" can be formed if a carbon atom is bonded to four

different groups. The isomeric structures are related in the same way

that an unsymmetrical object is related to its mirror image (Figure 2.2)

or the way a right hand is related to a left hand. Such components are

said to be "optically active" because they rotate the plane of polarized

light.

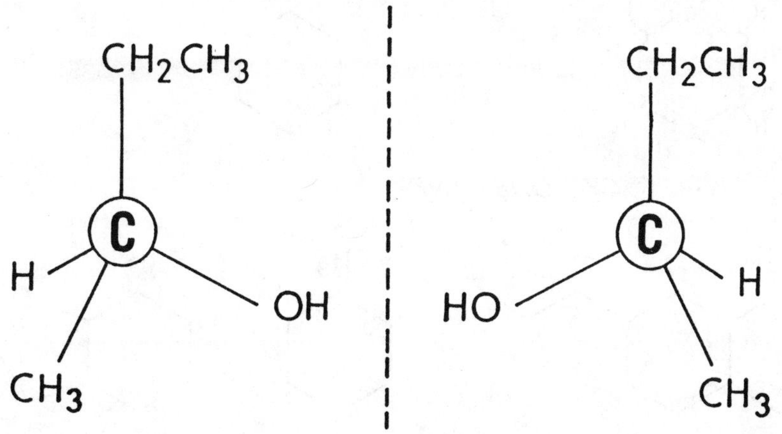

Figure 2.2:
Optical isomers. The spatial configurations of the two forms are
related as though the dotted line represents a mirror.

Rings ("naphthenes")

When the ends of a chain are linked together, two hydrogens are eliminated creating a ring structure of general formula C_nH_{2n}. The most common examples contain five or six carbon atoms in the ring (Figure 2.3). These rings can have one or more side chains and may be fused together. Alternative names for ring compounds of this type are "naphthenes" and "cycloparaffins".

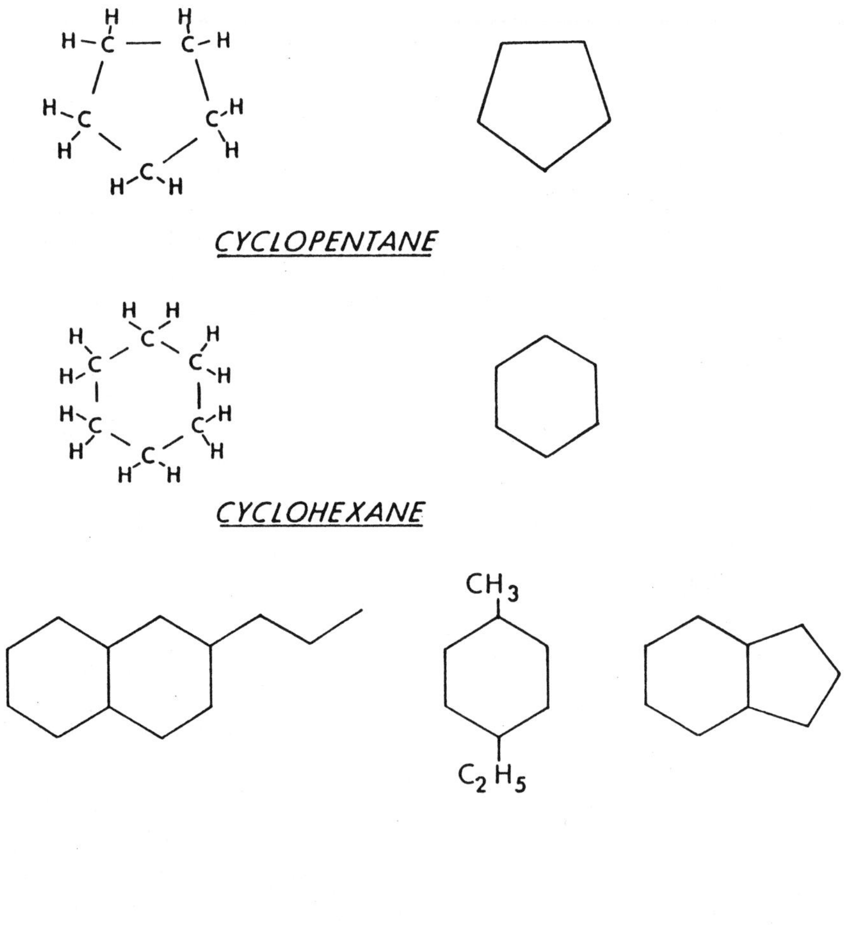

CYCLOPENTANE

CYCLOHEXANE

Figure 2.3:
Saturated ring compounds showing the arrangement
of carbon atoms, conventional representations,
and selected examples.

Unsaturated Compounds

In the compounds discussed so far the four bonds formed by each carbon have gone to four different atoms. These compounds are said to be "saturated." However, it is possible for a carbon atom to form more than one bond to another atom producing "unsaturated" compounds which contain double, or even triple, bonds:

$$
\begin{array}{ccc}
H & & H \\
\diagdown & & \diagup \\
& C = C & \\
\diagup & & \diagdown \\
H & & H
\end{array}
\qquad\qquad
H - C \equiv C - H
$$

ethylene   acetylene

It might seem that a multiple bond would strengthen a molecule, but the reverse is true. Multiple bonds are points of weakness and most unsaturated compounds (other than aromatics) are reactive and do not survive long in the subsurface.

Aromatics

A special situation arises when alternating double and single bonds occur in a six-membered ring. The simplest compound of this type is the hydrocarbon benzene. Unlike other compounds with double bonds benzene has considerable stability and shows little tendency to become saturated. A chemical bond causes an increase in electron density between the bonded atoms, and the density increase is greater for a double bond than for a single one. However, in benzene the electron density is spread out evenly, giving in effect six "one-and-a-half" bonds. It is this additional but uniform electron density distribution which is responsible for the enhanced stability. Since many simple benzene compounds of this type have pleasant fruity odors, benzene and related ring compounds are called "aromatics." Aromatic rings can be fused either to other aromatic

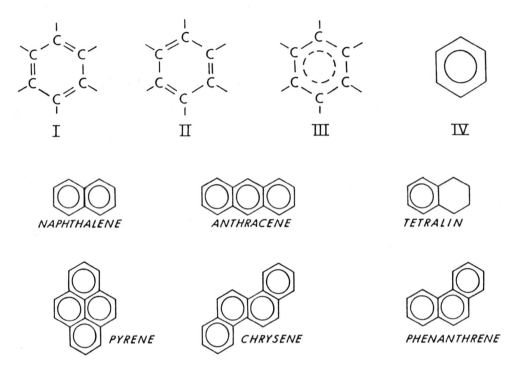

Figure 2.4:
Some conventional ways of representing benzene
(I to IV) and examples of fused ring compounds.
Note that in the tetralin structure the hexagon
without a circle represents a saturated,
non-aromatic ring.

rings ("polynuclear aromatics") or to saturated rings. Any of the rings

may have side chains (Figure 2.4).

Aromatic compounds have relatively low hydrogen contents and this

is well illustrated by comparing benezene (C_6H_6) with the C_6 normal

paraffin and the C_6 cyclic paraffin (naphthene) (Table 2.2).

COMPOUNDS WITH HETEROATOMS

In addition to carbon and hydrogen many molecules contain other

elements and these are called "heteroatoms". The heteroatoms of

interest in petroleum geochemistry are most commonly nitrogen, sulfur

and oxygen, which has led to the designation of these compounds as

COMPOUND TYPE	STRUCTURE	CONVENTIONAL REPRESENTATION	FORMULA	H/C
PARAFFIN			C_6H_{14}	2.33
NAPHTHENE			C_6H_{12}	2.00
AROMATIC			C_6H_6	1.00

Table 2.2:
Relative hydrogen contents of aromatic and
saturated hydrocarbons with six carbon atoms.

"NSO's". The addition of compounds with heteroatoms to the already large inventory of hydrocarbons containing only carbon and hydrogen produces an enormous number of different compounds. Fortunately there are relatively few ways in which the heteroatoms can be arranged. Each of the groupings is called a "functional group" and has specific chemical characteristics, so that they nearly always react in the same way regardless of the complexity of the organic compound to which they are attached. For example, the -COOH acid groups in the $C_{15}H_{31}$.COOH (palmitic acid) and $C_{17}H_{35}$.COOH (stearic acid) react similarly under most conditions. A list of the more important functional groups is given in Table 2.3.

Table 2.3:
Selected Common Functional Groups

Chemical Symbol	Name

Hydrocarbon Groups

$-CH_3$, $-C_2H_5$, $-C_3H_7$	methyl, ethyl, propyl
$-C_6H_5$	phenyl
$-R$, $-R'$, $-R''$	any alkyl group or predominantly hydrocarbon group
$-Ar$	any predominantly aromatic group

Oxygen-Bearing Groups

$-OH$	hydroxyl ("alcoholic" if linked to an aliphatic group; "phenolic" if linked to an aromatic group)
$-C\overset{O}{\underset{OH}{}}$	acid
$-C\overset{O}{\underset{OR}{}}$	ester
$C = O$	carbonyl
$R-\overset{O}{\underset{}{C}}-R'$	aldehyde if R' is H ketone if R' is an alkyl group
$R-O-R'$	ether
$-OCH_3$	methoxyl

Nitrogen-Bearing Groups

$-NH_2$	amino
$-C\equiv N$	nitrilo

Sulfur-Bearing Groups

$-SH$	mercaptan (thiol)
$R-S-R'$	sulfide (thioether)

Many compounds contain more than one functional group and frequently can include more than one heteroatom. Important examples for organic geochemistry include the amino acids since they are the units from which proteins are built. Amino acids, as the name suggests, contain both nitrogen in an amino group and oxygen in an acid group. Most of the large, naturally-occurring molecules (e.g. chlorophyll) have many attached groups and side chains. These may include one or more aliphatic groups such as methyl or ethyl, oxygenated groups like hydroxyl, methoxyl, and carboxyl, and nitrogen and sulfur either in heterocyclic rings or as amino or thiol groups.

ISOTOPES

Carbon has two stable isotopes, called carbon-12 and carbon-13, which differ slightly in mass but have essentially the same chemical properties. The relative amounts of these two isotopes vary somewhat in carbon-bearing materials, and these variations give useful geochemical information. It is normal to express the relative abundances in terms of a "δC^{13}" value. This value has units of parts per thousand, or "per mil" and is defined by the expression:

$$\delta C^{13} \ \text{\textperthousand} \ = \left[\frac{(C^{13}/C^{12})\ \text{SAMPLE} - (C^{13}/C^{12})\ \text{STANDARD}}{(C^{13}/C^{12})\ \text{STANDARD}} \right] X\ 1000$$

The most widely used standard is a belemnite carbonate from the Pee Dee formation, South Carolina, and is referred to as the "PDB standard". Other standards can easily be referred to it (Table 2.4). The δC^{13}

value of the standard is always zero per mil. Materials rich in C^{12} are said to be "light" and when they have more C^{12} than the standard they give negative δC^{13} values. Materials with more C^{13} than the standard have positive δC^{13} values.

Table 3.2:
Values for various carbon isotope
standards relative to the PBD standard
(Craig, 1957).

Name	Material	Value Relative to PDB ^{o}/oo
PDB	Belemnite from Pee Dee Formation	0.00 (Standard)
Nier	Solenhofen limestone	-0.64
NBS Reference Sample #20	Solenhofen limestone	-1.06
NBS Reference Sample #21	Graphite	-27.79
Basel	Ticino marble	+2.77
Wellington	Te Kuite limestone	-1.67
Stockholm	$BaCO_3$	-10.32

CHEMICAL REACTIONS

In general, we are concerned with reactions occurring at the earth's surface or in the subsurface at depths down to about 30,000 feet. Temperatures generally do not exceed a few hundred degrees C, though they may be higher locally in the vicinity of igneous intrusions. The rate of a chemical reaction increases with temperature but pressure seems to have only a minor role. Although near surface conditions may be oxidizing, the subsurface is almost entirely reducing. Thus, in general, the reactions of interest will be occurring at relatively low temperatures in a reducing environment and in the presence of water. Some types of reactions are sufficiently common that they have been given generalized names. These include the following:

Oxidation

Organic matter is unstable with respect to reaction with oxygen. Carbon dioxide and water are the ultimate products, although partial oxidations frequently occur.

Reduction

Reduction is the opposite of oxidation and can be thought of as removal of oxygen, or addition of hydrogen. An example is the reduction of an alcohol to a paraffin:

$$R-CH_2-OH + 2H = R-CH_3 + H_2O$$

Elimination

Some reactions occur with the loss of a small molecule like water, ammonia or carbon dioxide and are called "elimination reactions." Removal of the acid group from a fatty acid ("decarboxylation") is one example where the eliminated molecule is carbon dioxide and the product is a paraffin:

$$R'-CH_2-CH_2-COOH = R'-CH_2-CH_3 + CO_2.$$

Esterification is a reaction occurring between a carboxylic acid and an alcohol which results in the elimination of a molecule of water and the formation of an "ester" of general formula $R-C\underset{OR}{\overset{O}{<}}$. This is an example of a reversible reation which can proceed in either direction depending on temperature and concentrations.

$$R'-CH_2-C\overset{\displaystyle O}{\underset{OH}{\diagdown}} \quad + \quad HOR \quad = \quad R'-CH_2-C\overset{\displaystyle O}{\underset{OR}{\diagdown}} \quad + \quad H_2O$$

acid   alcohol   ester

Polymerization

Under the appropriate conditions small molecules can react with one another to build up much larger entities called polymers. The process is one of "polymerization". Many of the plastics in common use are made by polymerizing small organic molecules. For example, polyethylene is made by polymerizing ethylene and polystyrene by polymerizing styrene. Cellulose and protein are examples of naturally occurring biopolymers and are built up from large numbers of sugar or amino acid molecules, respectively. Polymerization is usually accompanied by elimination.

Pyrolysis

Pyrolysis is the thermally induced break-up of large molecules to give lower molecular weight products. During pyrolysis ring formation ("cyclization") and the development of aromatic structures from saturated ones ("aromatization") may occur.

EXPERIMENTAL METHODS

Most naturally-occurring carbonaceous materials are either very complex mixtures containing a large number of closely related compounds (such as crude oil), or are high molecular weight, polymeric materials which are not soluble in normal organic solvents. In most cases, the organic matter in rocks is present in low concentrations. It is only recently that techniques have been developed for separating and analyzing such organic materials, and it is this more than any other single factor that has contributed to the rapid development of organic geochemistry.

Separation

Soluble organic materials are removed from finely ground rocks by solvent extraction with benezene,* benzene-methanol, or other suitable solvent. Removal of the solvent by evaporation then leaves a "heavy hydrocarbon extract" containing compounds with fifteen or more carbon atoms and identified as the "C_{15+} extract." The remaining insoluble organic matter is called "kerogen" (by definition). It can be separated by dissolving away the rock matrix with mineral acids. Figure 2.5 presents a generalized scheme for separation and analysis.

Chromatography

Most of the analytical techniques available to organic chemists have been used in characterizing the solvent-soluble organics in geologic materials, but one group of techniques has been particularly important in separating the complex mixtures so often encountered. This falls in the general group of procedures known as "chromatography." Chromatography was first used by Tswett, a Russian botanist, to separate plant pigments (hence the "chroma") and relies on the fact that the different components in a mixture will migrate over the surface of a suitable substrate at different rates. If, for example, a mixture containing compounds A and B is poured into a vertical column packed with a suitable powdered solid (as shown schematically in Figure 2.6) and then pure solvent is contin-uously added the components in the mixture will be washed down the column. Separation occurs because the components move at different

---

* A recent OSHA ruling has reduced the acceptable benzene concentration in air to 1 ppm and many laboratories are in the process of changing to other solvent systems.

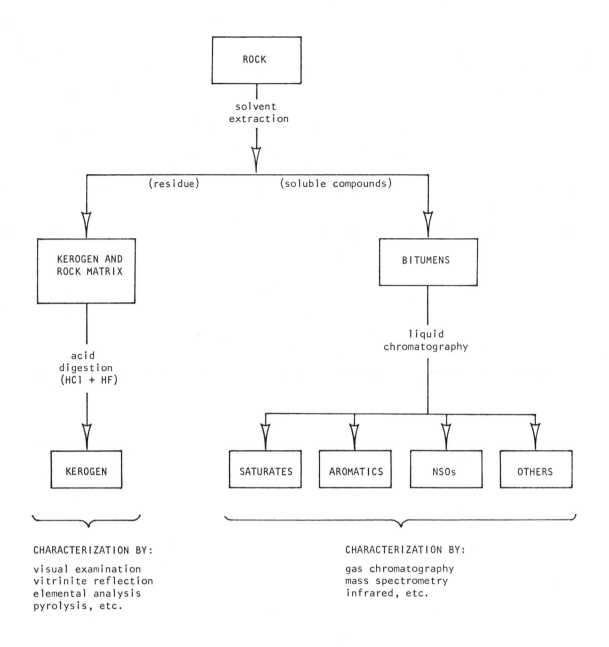

Figure 2.5: Generalized analytical scheme for rock samples.
Crude oils would be treated like "bitumens" after a preliminary
deasphaltening.

rates depending on their interaction with the packing - the greater the

interaction the slower they move. Liquid chromatography of this type, with

silica and alumina as the packing material, is widely used to separate

saturates, aromatics and NSO's.

Gas Chromatography (G.C.)

When a gas stream is used to carry the components over the packing

the technique is called "gas chromatography". It is widely used for separating complex volatile mixtures, such as the saturate fraction obtained by liquid chromatography. Samples as small as 1 μl are routinely analyzed but the amounts of separated components are too small to be collected. Instead, their presence and relative concentrations are registered as the output from some sensitive detector located at the end of the column. A typical gas chromogram is shown in Figure 2.7. The column temperature is steadily increased during the analysis to improve resolution of peaks and to permit the analysis of compounds with high boiling points. In general the lighter components come off the column first and many gas chromatographic separations approximate a boiling

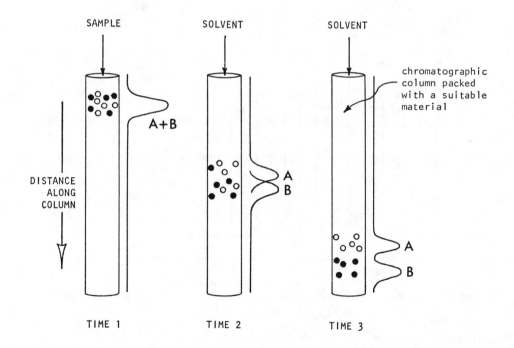

Figure 2.6
Schematic illustration of the principle of chromatographic separation. The compound A interacts more strongly with the column packing than compound B, so that A's rate of movement is impeded and it moves through the column more slowly than B.

21

point separation.

Gas chromatography is primarily a separation technique and it is sometimes difficult to identify the compound responsible for a given peak. The recent use of mass spectrometers (MS) in the role of gas chromatographic detectors has helped resolve this problem. In the mass spectrometer a compound is broken down into characteristic ionized fragments and identified on the basis of the masses and relative concentrations of these ions. The GC-MS instruments now being used provide an extremely powerful analytical tool.

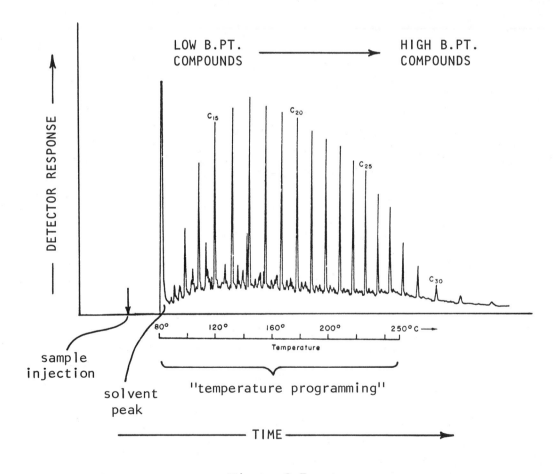

Figure 2.7:
Gas chromatogram. In this example the prominent peaks
are normal paraffins (labelled C_{15}, etc.).

Contamination during sampling

In designing a sampling program for a geochemical study it is important to recognize sources of contamination and to take steps to eliminate or minimize them. Contamination may occur during drilling and sample retrieval, shipping and storage, or during analysis. Diesel fuel is commonly used as a drilling mud additive, particularly when problems occur, and other organic materials (everything from wax to walnut shells) may also come in contact with samples. Samples of these potential contaminants can often help the laboratory geochemist and should be collected whenever possible. The act of removing rock fragments from a high temperature, high pressure environment to the suface poses potential problems, particularly in the loss of volatile hydrocarbons. Samples collected for analysis of these low moleculer weight hydrocarbons should be canned at the well site and shipped to the lab, sealed. Crude oil should be shipped in glass or metal containers, but never in plastic ones because the plasticizers dissolve in the oil and contaminate it.

REFERENCES

Banks, J.E., 1976. Naming organic compounds. A programmed introduction to organic chemistry (2nd Edition): Saunders 308 p.

Craig, H., 1957. Isotopic standards for carbon and oxygen and correction factors for mass-spectronmetric analysis of carbon dioxide: Geochim. Cosmochim. Acta, v. 12, p. 133-149.

Swain, F. M., 1970. Non-marine organic geochemistry (Chapter 1): Cambridge Univ. Press.

Weast, R. C. (Editor), 1970. Handbook of Chemistry and Physics (51st Edition): The Chemical Rubber Company, Cleveland p. C1-C52.

TOPIC 3

ORGANIC MATTER IN RECENT SEDIMENTS

INTRODUCTION

The organic matter now found in ancient rocks was incorporated from
the overlying water column at the time they were being deposited as
sediments. In the oceans organic matter occurs in true solution and as
particulate or colloidal material. The concentration of all three forms
is highest near surface but decreases with depth for the first few
hundred meters, and then remains approximately constant. Dissolved
organic matter is more than an order or magnitude more abundant than
particulate organic matter. As clay particles settle through the water
they adsorb some organic matter from solution and this is then incorporated into the sediments. Particulate matter settles to the bottom
and becomes part of the sediments, while the colloidal organic matter
first flocculates and then settles (Figure 3.1). As the organic matter
sinks through the water column bacteria and other organisms remove any
of the organic compounds which they can metabolize, and at the water-
sediment interface bottom feeders further screen the organic matter so
that only relatively resistant organic material is incorporated into the
sediments. The only exceptions to this are the organic compounds which
are protected in some way. For example shells are held together by
proteinaceous material which is protected from degradation by the
enclosing carbonate.

Both the amount and type of the organic material reaching the
sediments is controlled, at least in part, by the depositional environment and can be related to the productivity of the overlying waters, the

24

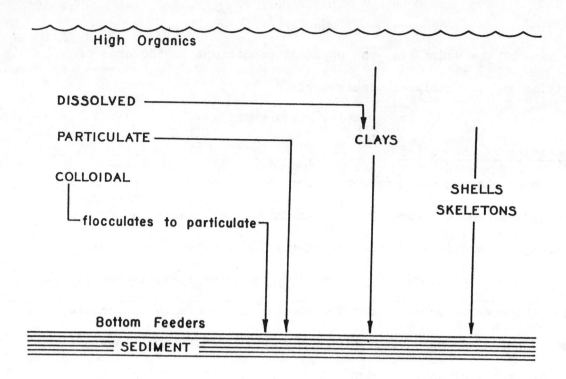

Figure 3.1: Deposition of organic matter in aqueous environments.

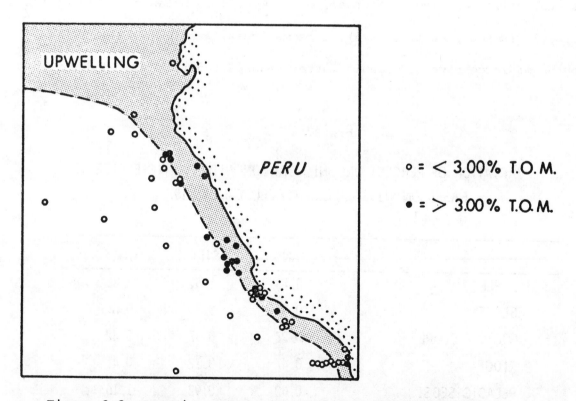

Figure 3.2: Organic matter contents of sediments in the Pacific, West of Peru showing high values under the productive Humboldt Current and low values elsewhere. Data from Rosato, et al. (1975).

grain size of the sediments, the physical conditions in the area of
deposition and the sediment mineralogy.

SEDIMENTARY ENVIRONMENTS

Productivity of the waters

Much of the organic matter in ocean waters comes from the indige-
nous organisms so that more organic material is available in areas of
high marine productivity. This is often reflected by higher organic
matter contents in the sediments underlying highly productive surface
waters. Figure 3.2 shows values for the organic contents of sediments
deposited below the very productive Humboldt Current which flows north-
ward along the west coast of South America. All the samples with more
than 3% organic carbon are under, or close to, the highly productive
waters, while the samples farther to the west have organic carbon contents
below 3% (Rosato, et al., 1975). In some areas, particularly near the
mouths of large rivers, organic matter transparted from the land surface

Table 3.1:

VARIATION OF PERCENT ORGANIC CARBON WITH PARTICLE SIZE
(BASED ON 911 SAMPLES) (YEMEL' YANOV, 1975)

	SAND	SILT	CLAY
SHELF	0.73	1.35	2.86
SLOPE	0.44	0.83	0.85
FOOT OF SLOPE	0.27	0.37	0.42
RIDGE	0.47	0.12	0.41
PELAGIC SEDS.	0.62	0.92	1.16

may be quantitatively important.

Grain size

There is a well-documented relationship between organic matter content and grain size in recent sediments (Table 3.1) and this trend is preserved in ancient rocks. Three main processes operated to produce this distribution:

a) Wave and current action winnows the clay-sized materials and small particles of low density organic material out of the sands and deposits them together in quieter waters.

b) The high energy environments (such as beaches, bars, etc.) which characterize sand deposition are often oxygenated, and organic matter is not stable in an oxidizing environment.

c) Settling clay particles, because of their large surface areas, may adsorb some types of organic materials from solution and transfer them into the sediments.

Physical conditions

Organic matter is not stable under oxidizing conditions but can be preserved when the environment is reducing. Oxidizing conditions are often indicated by the nature of the inorganic constituents, for example, by the presence of iron oxides. Since these give the color to "red beds" rocks of this type generally have low organic matter contents and therefore are usually excluded from consideration as source rocks.

Conditions below the surface of sediments rapidly become reducing and aerobic bacteria can no longer oxidize organic matter. Anaerobic bacteria continue to function using the oxygen from sulfate but their effect diminishes as depth increases.

Rock type

Sands usually have low organic matter contents because of their high energy, oxidizing environments of deposition, while shales de-

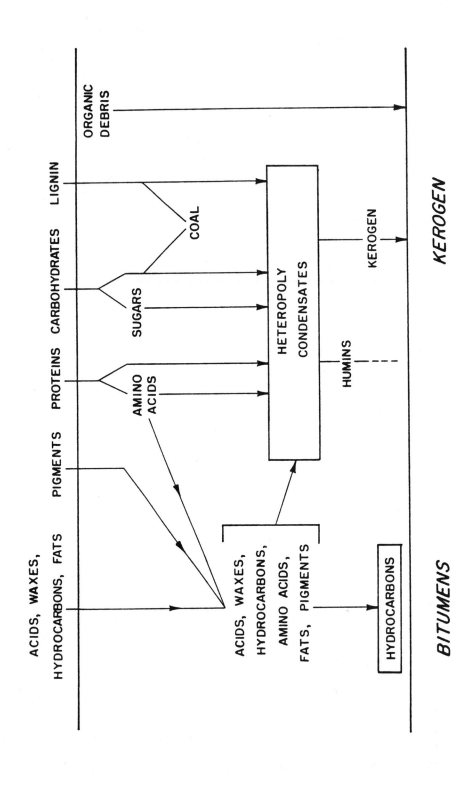

Figure 3.4: The fate of organic matter in recent sediments. The major types of organic materials made by living organisms are given above the top line and the materials found in ancient sediments (rocks) summarized below the bottom line. In between the major changes produced in the water column and in recent sediments are summarized.

posited under reducing conditions have high percentages of organic

matter.   Carbonates usually are intermediate (Figure 3.3).

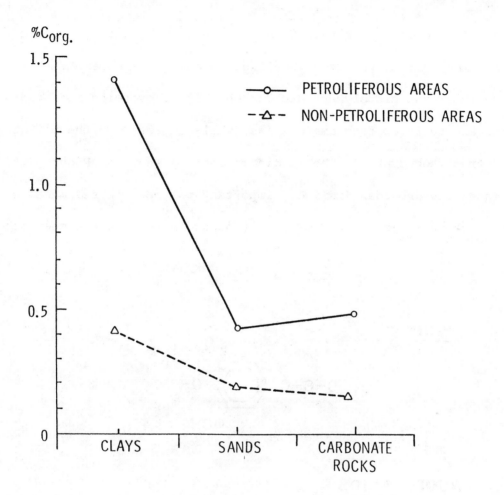

Figure 3.3:
Average content of organic carbon in sedimentary rocks
(Data: Ronov, 1958 25,742 Samples (1105 composites))

ORGANIC COMPOUNDS

The total amount of organic material incorporated into sediments is

important, but so is its chemical composition. The nature of the organic

compounds which become part of the sediment depends on the type of

organisms living in the overlying water column and on the relative contribution of transported organic debris from terrestrial and other environments. Several classes of organic compounds are preserved in recent sediments and these are described below and listed in Figure 3.4.

Biopolymers

Much of the organic matter is high molecular weight, polymeric material made up from repeated, small units called "monomers". Among the more important biopolymers are carbohydrates, proteins and lignin. Proteins are built up from amino acids, while sugars form the building blocks for carbohydrates. The lignin monomer is shown in Figure 3.5. Humic materials are quantitatively important polymers in sediments but do not occur in living organisms. They occur in soils where they are

Figure 3.5:

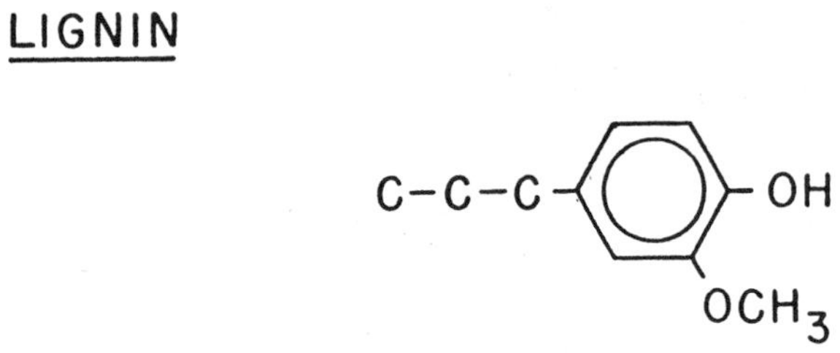

LIGNIN

C–C–C–⟨◯⟩–OH
OCH_3

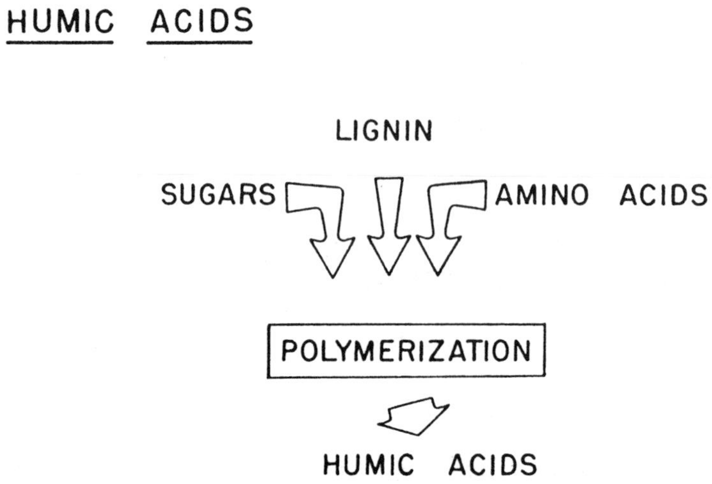

HUMIC ACIDS

LIGNIN
SUGARS   AMINO ACIDS

POLYMERIZATION

HUMIC ACIDS

responsible for the dark color and are synthesized from many smaller molecules including amino acids, lignin monomers and sugars.

Bitumens

Although biopolymers are not soluble in normal organic solvents like benzene and methanol, many other types of compounds are. This important group is called the "bitumens," "lipids," or simply the "extractables." It includes many compound types which are important in considerations of petroleum generation and migration.

Fatty acids form a major part of the lipids in many organisms and have sufficient stability to persist for geologically long time intervals. Most fatty acids are saturated straight chain compounds; branched chain and unsaturated acids are less common. Fatty acids from organisms and recent sediments show a marked predominance of compounds with even numbers of carbon atoms, for example the C_{16} and C_{18} acids are frequently the most abundant and are always present in greater amounts than the C_{17} acid. This even predominance can be quantified by using ratios such as $(nC_{16} + nC_{18})/(2 \times nC_{17})$. Values greater than 10 are not uncommon in recent sediments, but decrease with age and usually are in the range 1 to 2 for ancient sediments. Fatty acids are believed to be important precursors for petroleum n-paraffins.

Waxes are a major component of the lipids from terrestrial plants since they form the protective layer on leaves, etc. They are esters of a long chain acid and a long chain monohydroxy alcohol (Figure 3.6).

Hydrocarbons are an important contributor to the bitumens and may be saturated, unsaturated, cyclic, or aromatic. In living systems there are negligible amounts of aromatics, and the unsaturated hydrocarbons are much more abundant than the saturated ones. The straight chain

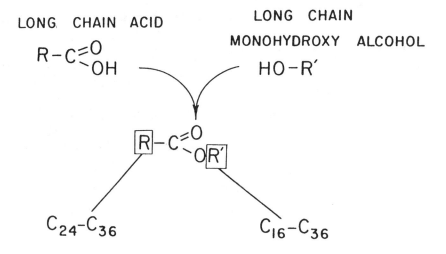

LONG CHAIN ACID

LONG CHAIN
MONOHYDROXY ALCOHOL

Figure 3.6: Waxes.
R is generally in the range C_{24} to C_{36} and
R' in the range C_{16} to C_{36}.

paraffins are the best studied because of their close relationship to the hydrocarbons in crude oils. In living systems and recent sediments the C_3 - C_{10} compounds are almost completely absent. Primitive plants, bacteria, and algae generally show a maximum between nC_{17} and nC_{21}, whereas the more specialized terrestrial plants have a maximum around nC_{29}. In both living systems and recent sediments the paraffins containing an odd number of carbon atoms are more abundant than paraffins with even carbon numbers (note that this is the reverse of the trend observed for the fatty acids). A quantitative expression for the odd-even preference, called the Carbon Preference Index (CPI), is defined by an expression such as

$$CPI = \frac{1}{2}\left[\frac{\Sigma nC_{17} \text{ TO } nC_{31}}{\Sigma nC_{16} \text{ TO } nC_{30}} + \frac{\Sigma nC_{17} \text{ TO } nC_{31}}{\Sigma nC_{18} \text{ TO } nC_{32}}\right]$$

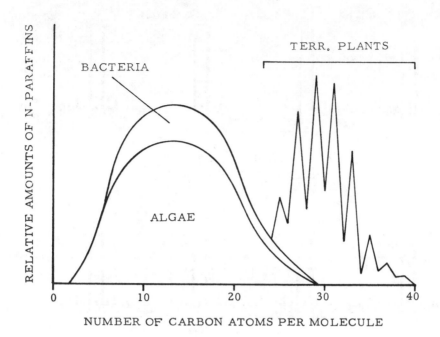

Figure 3.7:
Chain length distribution for n-paraffins from
various types of organic matter (Modified from
Lijmbach, 1975).

If even- and odd-numbered paraffins are equally abundant the value of
the CPI will be 1, but if the odd-numbered compounds predominate the CPI
will be greater than 1. Values up to 4 or 5 have been reported for
hydrocarbons from living systems and recent sediments, while values
closer to 1 are more typical of the bitumens in ancient rocks.

Many types of branched hydrocarbons occur in the bitumens from
recent sediments but the isoprenoids are particularly important because
they are relatively abundant and are directly related to biological
precursors. In principle they are derived from repeated isoprene units
which leads to a chain with a branch on every fifth carbon atom. (In an
alternative nomenclature isoprenoid compounds are called terpanes.) The
most important hydrocarbons which have the characteristic branching are
"pristane" (C_{19}) and "phytane" (C_{20}). These are present in living

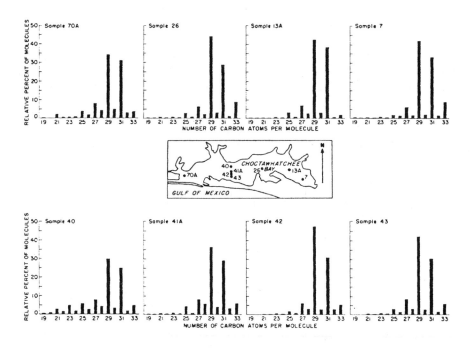

Figure 3.8:
Distribution of n-paraffins C_{19}-C_{33} in upper layer
of estuarine mud, Choctawhatchee Bay, Florida (Palacas et al.,
1972). The input here is almost solely terrestrial.

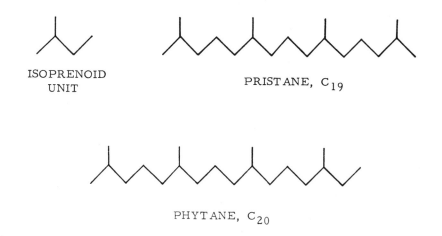

ISOPRENOID
UNIT

PRISTANE, C_{19}

PHYTANE, C_{20}

Figure 3.9:
Isoprenoid hydrocarbons showing the characteristic
branching on every fifth carbon atom in the chain.

systems, recent sediments, ancient sediments, and crude oil and their

relative abundances have been used to infer environments of deposition.

The pristane/nC_{17} and phytane/nC_{18} ratios have also been used as environmental indicators (Lijmbach, 1975).

Cyclic hydrocarbons (naphthenes) are based on fused five- and six-membered rings. The characteristic structure of the steranes contain

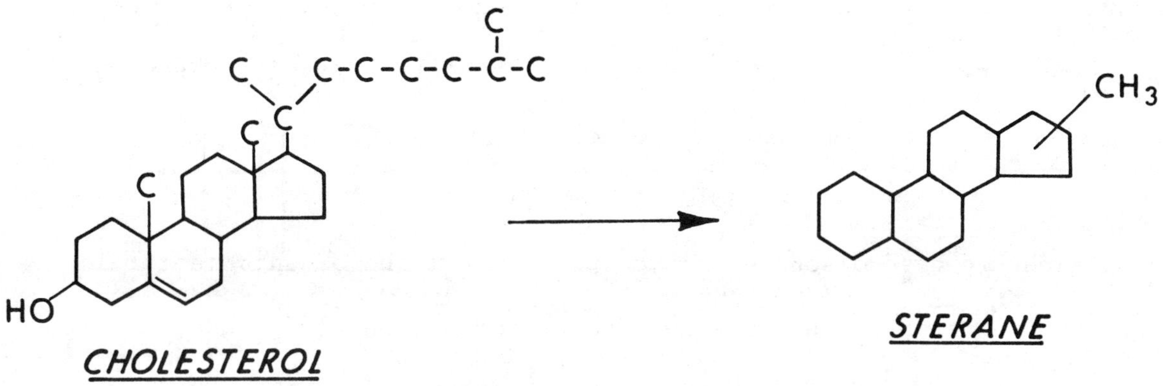

CHOLESTEROL → STERANE

Figure 3.10:

PORPHYRIN

ISOPRENOID

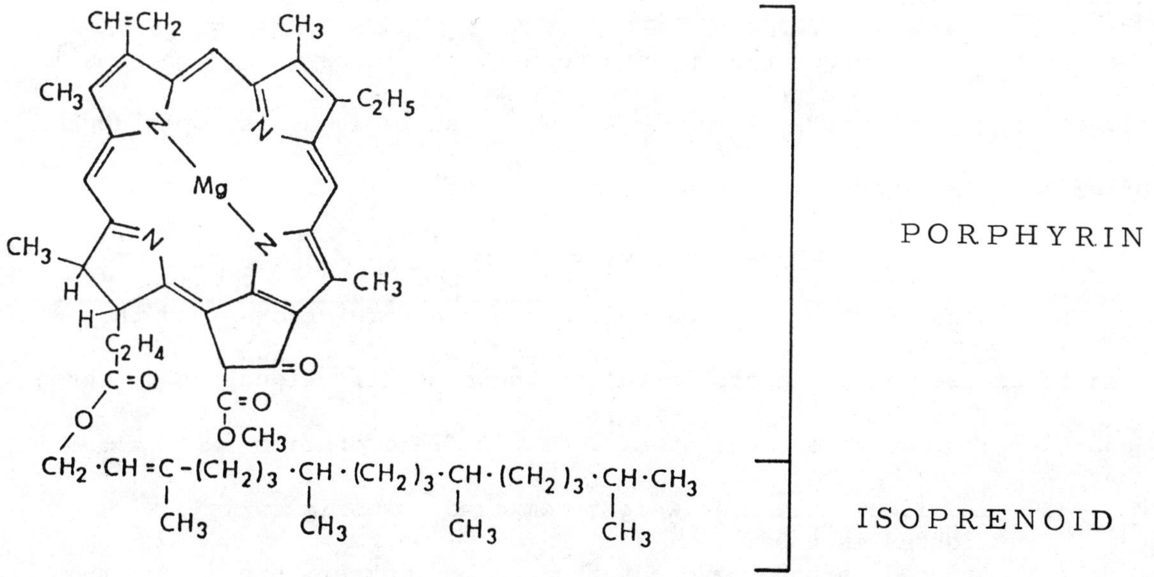

Figure 3.11: Chlorophyll

three 6-membered rings and one 5-membered ring and appears to be derived from biochemically-produced steroids such as cholesterol (Figure 3.10).

Other components of bitumens include fats, alcohols and pigments. The fats are important because they hydrolyze to give acids, and the alcohols can lead to hydrocarbons by reduction. Chlorophyll is quantitatively important and is a source of isoprenoids and porphyrins (Figure 3.11) while the red pigment, carotene, is thought to provide a source for many of the smaller aromatic molecules.

ISOTOPIC COMPOSITION

So far we have discussed the composition of the organic matter in sediments in terms of the relative amounts of various compounds, but it is also necessary to consider the relative amounts of the two stable isotopes of carbon, since there are significant variations.

The isotopic composition of plant material is influenced by the isotopic composition of the source material. Carbon dioxide in the atmosphere is richer in C^{12} (i.e., "lighter") than the carbon dioxide in sea water. Thus terrestrial plants tend to be richer in C^{12} than marine plants (such as algae). Also, in sea water there is an isotopic equilibrium between carbon dioxide and bicarbonate:

$$C^{13}O_2 + HC^{12}O_3^- \rightleftharpoons C^{12}O_2 + HC^{13}O_3^-$$

Reactions of this sort are called "exchange reactions" and they lead to changes in the isotopic ratios known as "fractionation." Three general rules govern the extent of isotopic fractionation.

(1) The lower molecular weight compound is enriched in the lighter isotope.
(2) Reduced carbon compounds tend to be lighter than oxidized carbon compounds.
(3) The amount of fractionation decreases with increasing temperature.

Therefore, in the equilibrium reaction given above the carbon dioxide (molecular weight 44) is enriched in C^{12} relative to the bicarbonate (molecular weight 61), but the enrichment decreases as temperature increases.

Additional isotopic fractionation is produced because living systems incorporate the two carbon isotopes at different rates leading to a "kinetic isotope effect." The net result of all these processes is to produce terrestrial organic matter that is richer in C^{12} than marine organic matter, so that with increasing distance from shore the organic matter in sediments gets heavier (Sackett and Thompson, 1963; Gearing et al., 1977) (Figure 3.12).

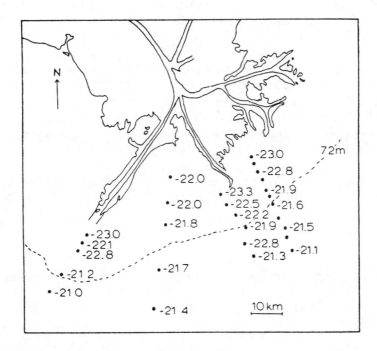

Figure 3.12:
Isotopic composition of organic carbon in the recent sediments being deposited off the Mississippi delta. Note that isotopic values get "heavier" with increasing distance from the land.

REFERENCES

Bordovskiy, I. K., 1965. Accumulation and transformation of organic substances in marine sediments: Marine Geol., v. 3, p. 3-114.

Degens, E. T. and Mopper, K., 1976. Factors controlling the distribution and early diagenesis of organic material in marine sediments: Chapter 31 in "Chemical Oceanography" (Ed. by J. P. Riley and R. Chester) Academic Press Vol. 6 (2nd Ed), p. 59-113.

Gearing, P., Plucker, F. E. and Parker, P. L., 1977. Organic carbon stable isotope ratios of continental margin sediments: Marine Chem., v. 5, p. 251-266.

Lijmbach, G. W. M., 1975. On the origin of petroleum: Proc. 9th World Petrol. Conf., Applied Science Publishers, London. v. 2 p. 357-369.

Palacas, J. G., Love, A. H. and Gerrild, P. M., 1972. Hydrocarbons in esturine sediments of Choctawhachee Bay, Florida, and their implications for genesis of petroleum: Bull. AAPG, v. 56, p. 1402-1418.

Ronov, A. B., 1958. Organic carbon in sedimentary rocks (in relation to the presence of petroleum): Geochemistry (Trans. from Russian), v. 5, p. 510-536.

Rosato, V. J., Kulm, L. D. and Derks, P. S., 1975. Surface sediments of the Nazca Plate: Pacific Sci., v. 29, p. 117-130.

Sackett, W. M. and Thompson, R. R., 1963. Isotopic organic carbon composition of recent continental derived sediments of eastern Gulf Coast, Gulf of Mexico: Bull. AAPG, v. 47, p. 525-528.

Yemel 'Yanov, Ye. M., 1975. Organic carbon in Atlantic sediments: Doklady Acad. Sci. USSR (English translation) v. 220, p. 220-223.

TOPIC 4

PETROLEUM GENERATION

INTRODUCTION

Average shales contain approximately 1 percent organic matter

(Hunt, 1972). About 90 percent of this is the high molecular weight,

insoluble, polymeric material called kerogen and the rest is solvent

soluble and is the bitumen fraction (Figure 4.1). The bitumens change

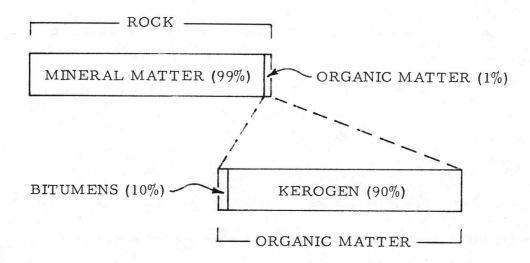

Figure 4.1:
Kerogen and bitumen in average shales

in amount and composition in response to changing biological and physical

conditions. This leads to important differences in chemical composition

between the bitumens in recent and ancient sediments, particularly for

the smaller molecules (Table 4.1). The classification of sediments into

"recent" or "ancient" is a matter of convenience only and does not

reflect any sharp boundary -- rather there is a gradual transition from

39

one to the other. Similarly with the organic matter there is a gradual change from the composition found in recent sediments to that found in the older rocks. In this topic we will be concerned with the systematic variations in the composition of the organic matter and with the factors which cause the changes.

Table 4.1:
Concentrations of low molecular weight hydrocarbons
(Dunton and Hunt, 1962) and light aromatics
(Erdman et al., 1958) in recent and ancient sediments.

	Recent	Ancient
Low mol. wt. hydrocarbons (C_4-C_8)	< 0.002 ppm*	2-1000 ppm
Light aromatics (Benzene + naphthalenes)	< 0.1 ppm*	1-25 ppm

*below limit of detection

The biochemical processes that operate in living systems generate a wide variety of organic compounds. Some of these are not stable and do not survive when the organism dies, but others are very stable and survive deep into the sedimentary sequence. Thus the organic matter which arrives in a sediment from the overlying water column contains many compounds with varying stabilities and these react at different rates in response to changing conditions.

Thermodynamic calculations show that in a reducing environment, methane and carbon are the stable end products produced by the influence of increased temperature on all types of organic matter. This implies a redistribution of hydrogen to give one product rich in hydrogen (CH_4) and one with no hydrogen (C). For example, the nC_{21} paraffin would ultimately redistribute its hydrogen thus:

40

$$C_{21}H_{44} \longrightarrow 11CH_4 + 10C.$$

With the increasing temperature that accompanies increasing depth of burial the kerogen gets progressively richer in carbon by becoming more condensed and developing more aromatic structures while the extractable fraction gets richer in hydrogen as the amount of paraffins increases. This redistribution of hydrogen can be summarized in the following scheme:

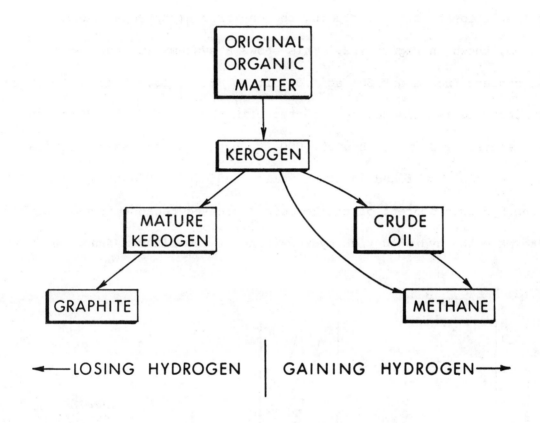

The quantity of bitumens increases because they are formed from the kerogen. We will see later that part of the bitumen fraction migrates and accumulates as crude oil so that the formation of bitumens from kerogen is the process of petroleum generation. Since the kerogen accounts for the bulk of the organic matter in rocks and has a major role in forming bitumens, changes in kerogen composition with depth provide an important indication of the progress of generation.

CHANGES IN KEROGEN COMPOSITION

Elemental composition

Various graphical representations have been developed for displaying changes in the chemical composition of kerogen as a function of depth. The simplest one is a plot of %C, %H, etc., against depth. LaPlante (1974) has done this for kerogens separated from shales in several Gulf Coast wells, including the one from South Pecan Lake, Louisiana, shown in Figure 4.2. This shows a systematic increase in carbon content from about 68% at 4,000 ft to 82% at 15,000 ft. The carbon increase is balanced by a decrease in oxygen content. Hydrogen content increases down to about 10,000 ft. but then decreases. Similar hydrogen trends were found in other wells studied. LaPlante suggested that the hydrogen content of the kerogen decreased because it was being used to make the hydrogen-rich hydrocarbons of the bitumen fraction. The

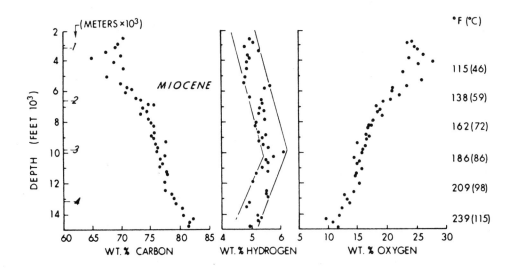

Figure 4.2:
Carbonization of kerogen with depth, South Pecan Lake field,
Cameron Parish, Louisiana. (LaPlante, 1974).

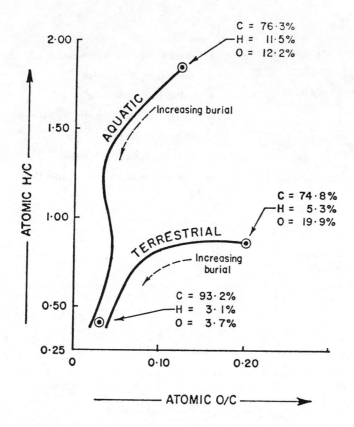

Figure 4.3a: Generalized diagenetic tracks
showing changes in atomic O/C and H/C ratios with increasing
depth for Type I, aquatic (mainly algal) and Type III,
terrestrial (mainly woody) types of organic matter. Selected
values for elemental weight percentages are given to
facilitate comparison with Figure 4.2.

occurrence of wet gas at 11-13,000 ft. in the South Pecan Lake well

supports this interpretation.

Percentages of carbon, hydrogen and oxygen can be plotted on C-H-O

ternary diagrams, but a more convenient alternative method has been

developed by coal researchers. They have used plots of atomic H/C

ratios versus atomic O/C ratios. Such diagrams are suitable for use

with kerogen data and have been applied successfully, particularly by

French geochemists. Tissot et al. (1974) have called the path defined

by H/C versus O/C values for kerogens from increasing depths a "dia-

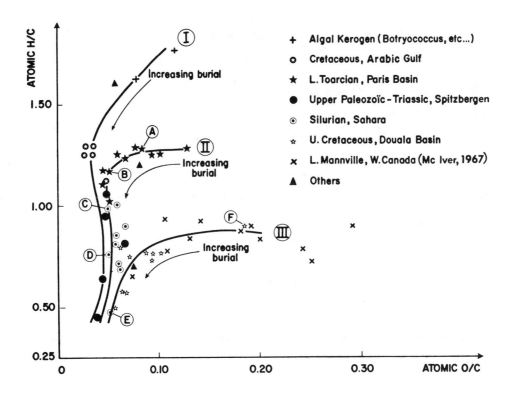

Figure 4.3b: Examples of kerogen evolution paths. Path I
includes algal kerogen and excellent source rocks from Middle
East; path II includes good source rocks from North Africa and
other basins; path III corresponds to less oil-productive
organic matter, but may include gas source rocks. Evolution of
kerogen composition with depth is marked by arrow along each
particular path. (Tissot, et al., 1974).

genetic track." Different types of organic matter follow different

tracks depending on their original elemental compositon (Figure 4.3 a & b).

Functional groups

The methods used for representing kerogen compositions in terms of

elemental percentage of carbon, hydrogen and oxygen give no indication

of the way in which heteroatoms are distributed between various func-

tional groups; for example, is the oxygen present as acidic groups,

alcoholic groups, phenolic groups, etc.? The detailed analysis of

functional groups is difficult, time consuming and not well suited to

44

the routine analysis of large numbers of samples. The original studies were carried out on coals of various ranks and diagrams such as Figure 4.4 were developed. In this figure the shaded area corresponds to the carbon percentages commonly measured for kerogens. The data show that oxygen in acid groups is eliminated early in diagenesis, but phenolic-OH groups survive through most of the range. Loss of carbonyl oxygen is intermediate.

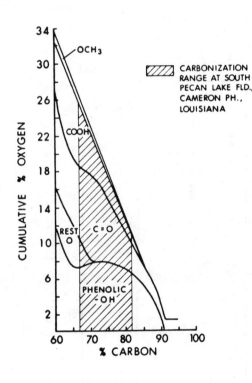

Figure 4.4:
Elimination of oxygen-bearing functional groups. Data from Blom (1961) with additions by LaPlante (1974).

CHANGES IN PHYSICAL CHARACTERISTICS OF KEROGEN

Kerogen color

As kerogen is heated to higher and higher temperatures in the subsurface it begins to generate petroleum compounds and its own chemical composition changes in a process called "maturation." Changes in the chemical composition of the kerogen are reflected in its physical appearance. Shallow (immature) samples generally contain kerogen which is light yellow in color but with increasing depth there is a progressive darkening through golden yellow to brown and finally to black. Several schemes have been introduced for rating the color intensities of kerogens, but the most widely used system employs a scale from 1 (for light golden yellow) to 5 (for black). Systematic trends with depth are seen in many basins, and Figure 4.5 shows how color darkens with depth of burial for the Toarcian shales of the Paris basin and the Logbaba formation of the Douala basin.

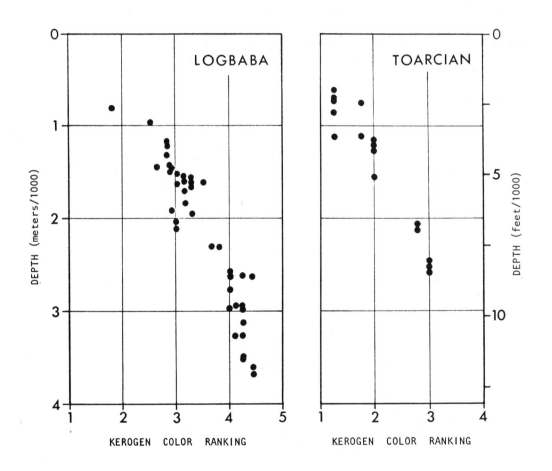

Figure 4.5: Darkening of kerogen color with depth in the Toarcian shales of the Paris basin and the Logbaba series of the Douala basin (after Correia, 1971).

Vitrinite reflectance

As kerogen matures the percentage carbon in the kerogen increases and it becomes "shinier" and reflects more of the light falling on it. Thus the "reflectance" (R_O) can be a useful measure of organic matter maturity. For comparative purposes it is necessary to select only one type of organic matter and vitrinite (a woody material) is universally used because of its uniform composition and predictable (and well-documented) behaviour (Figure 4.6).

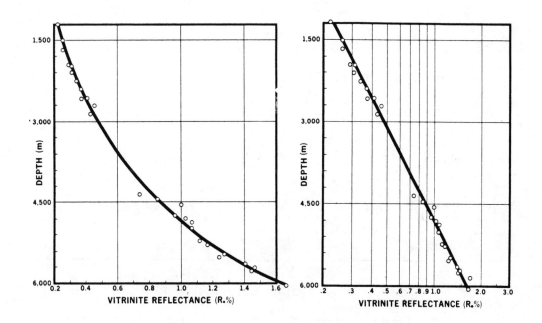

Figure 4.6: Data from Dow (1977).
A: Kerogen maturation profile of a typical Louisiana Gulf Coast well showing reflectance increase with depth. Both depth and average R_O values are plotted on linear scales.

B: Same maturation profile as shown in A but with R_O values plotted on a logarithmic scale. This is the simplest case of a continuously subsiding basin.

GENERATION OF HYDROCARBONS

Since kerogen accounts for over ninety percent of the organic matter in shales, small changes in its composition can produce large changes in the nature of the lower molecular weight, solvent extractable

bitumens that are generated from it. The trend of increasing carbon

content in the kerogen with progressive burial is balanced by an increasing

hydrogen content in the bitumens, and they become more paraffinic and

lower in molecular weight as thermal maturation proceeds. The net

effect is to transfer hydrogen from the kerogen to the bitumens (Figure 4.7).

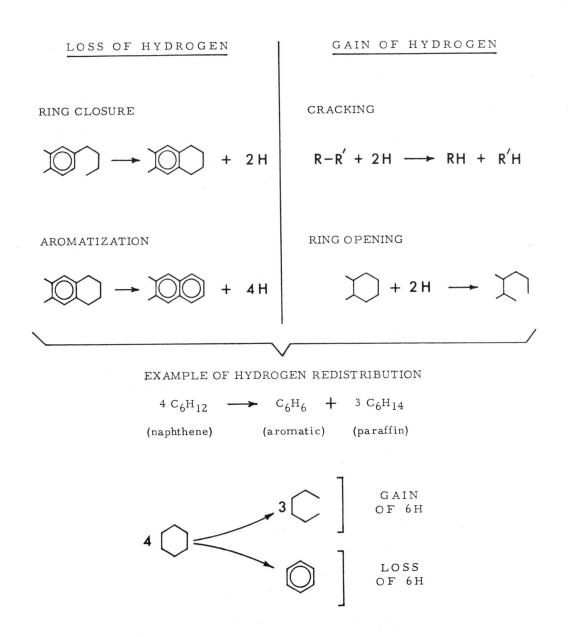

Figure 4.7: Schematic representation of the redistribution of hydrogen during the diagensis of organic matter.

The formation of lower molecular weight hydrocarbons is of particular importance because this is the process of petroleum generation. The C_2-C_{10} hydrocarbons, which are absent in living organisms and recent sediments but are such an important part of many crude oils, must be made from some other carbonaceous material. Kerogen is the most likely source because of its availability and its response to changing temperature.

FACTORS INVOLVED IN DIAGENESIS

The chemical compositions of kerogen and bitumens change systematically as depth increases. This is also the direction of increasing time, temperature and pressure and these factors must be evaluated for their influence on petroleum generation. Radioactivity, bacteria and catalysts have also been postulated to have a role in the maturation of organic matter but, except for the bacterial generation of some shallow dry gas, their influence seems to be minor. In most cases pressure also seems to be a minor factor, and in the following sections emphasis is placed on the effects of temperature and time.

Temperature

Temperature is the most important parameter in controlling the generation of petroleum. For each $10^{\circ}C$ ($18^{\circ}F$) rise in temperature the rate of reaction approximately doubles, so that although reaction rates are fairly slow near surface they increase rapidly with depth. This was demonstrated very clearly by Philippi (1965) in a classic study of Neogene sediments from the Los Angeles and Ventura basins of Calfornia. For rocks of different depths he determined the amount of organic matter and the amount and composition of the solvent extractable bitumens in the C_{15+} range. His curves showing the ratio (extractable C_{15+} hydrocarbons/

total organic carbon) as a function of sample depth are reproduced here as Figure 4.8. Both show a rather low initial ratio which increases rapidly at 10,000 ft. in the Los Angeles basin and at 16,000 ft. in the Ventura basin. What makes this study so remarkable is that the increase in the ratio occurs at the same temperature -- 120°C -- in both basins. This is compelling evidence that the generation of the extractable hydrocarbons is controlled by temperature; other factors, such as pressure, have only a minor role, if indeed they have a role at all. For example, in the Los Angeles basin the pressure is 5000 psi at the onset of generation but it is 60 percent higher at 8000 psi when generation starts in the Ventura basin.

LOS ANGELES BASIN   VENTURA BASIN

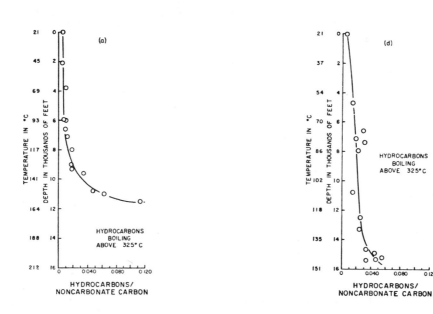

Figure 4.8:
Variation of the (hydrocarbon/noncarbonate carbon) ratio with
depth and age in shales from the Los Angeles and Ventura basins,
California (Philippi, 1965).

Many of the early studies in geochemistry were concerned with establishing the organic matter content of different rock types. Subsequent studies determined the areal extent of various organic matter levels and were in a sense "horzontal studies." Philippi's work in the California basins ushered in the era of "vertical studies" which placed the main emphasis on changes undergone by organic matter as it is buried. However, there are several inherent limitations in single-well studies of this sort. The organic matter in the lithologic units penetrated usually represents many different types of environments, and so variations due to organic matter type are superimposed on the trends produced by increasing temperature. Since the deeper rocks are also the oldest ones the effects of time cannot be distinguished from those of temperature. Some of these problems can be eliminated by studying a single rock unit which has been buried to different depths in different areas. For example, in a simple basin a given lithologic unit will be buried more deeply in the center than on the flanks. Tissot et al., (1971) selected the Jurassic Toarcian shale of the Paris Basin, France for a study of this type. This marine, organic rich shale has been buried to 7,500 ft. in the deepest part of the basin and outcrops on the flanks, and many wells have been drilled through it so that samples were available from different depths for geochemical analysis. The findings are summarized in Figure 4.9 and show that the (extract/total organic carbon) ratio increases with increasing depth just as in the Californian basins. Also as depth increases normal paraffins show diminshing odd-even preference and trend towards increasing abundances in the shorter chains. The 1- and 2-ring naphthenes increase in abundance at the expense of the multi-

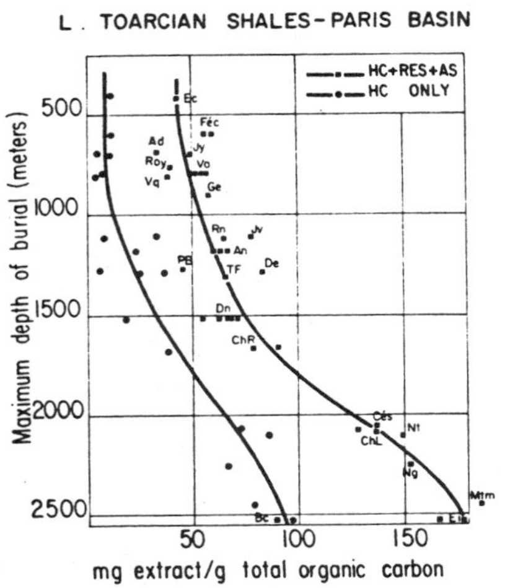

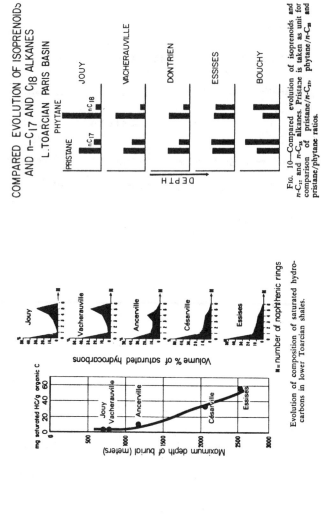

COMPARED EVOLUTION OF ISOPRENOIDS
AND n–C₁₇ AND C₁₈ ALKANES
L.TOARCIAN PARIS BASIN

JOUY

VACHERAUVILLE

DONTRIEN

ESSISES

BOUCHY

PRISTANE PHYTANE
nC17 nC18

DEPTH

Fig. 10—Compared evolution of isoprenoids and n-C₁₇ and n-C₁₈ alkanes. Pristane is taken as unit for comparison of pristane/n-C₁₇, phytane/n-C₁₈ and pristane/phytane ratios.

Evolution of composition of saturated hydro-carbons in lower Toarcian shales.

Jouy
Vacherauville
Ancerville
Césarville
Essises

Volume % of saturated hydrocarbons

N= number of naphthenic rings

mg saturated HC/g organic C
0 20 40 60

Maximum depth of burial (meters)
500
1000
1500
2000
2500
3000

Jouy
Vacherauville
Ancerville
Césarville
Essises

L. TOARCIAN SHALES–PARIS BASIN

HC+RES+AS
HC ONLY

Maximum depth of burial (meters)
500
1000
1500
2000
2500

mg extract/g total organic carbon
50 100 150

Fig. 5—Variation in amount of hydrocarbons ("HC only") and total chloroform extract (HC + RES + AS) as function of depth in lower Toarcian shales, Paris basin. Letters refer to origin of sample.

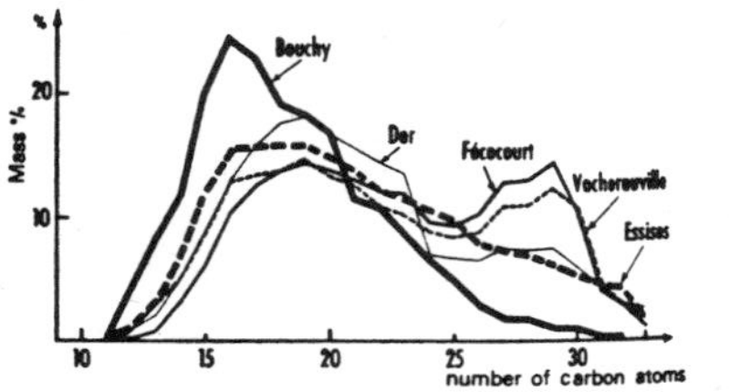

Mass %
20
10

10 15 20 25 30
number of carbon atoms

Bouchy
Ber
Hericourt
Vacherauville
Essises

Fig. 15—Distribution of aromatic hydrocarbons according to number of carbon atoms per molecule. Note regular decrease of deep samples, whereas shallow ones show second mode around C₂₁-C₂₅.

Figure 4.9: Organic geochemical data for the L. Toarcian shales of the Paris Basin.
(Tissot et al., 1971).

ringed naphthenes, and the relative amount of pristane and phytane decrease compared to the nC_{17} and nC_{18} normal paraffins.

A deep well drilled through the Logbaba series of the Douala basin, Cameroons provided an excellent opportunity to monitor changes in organic matter composition through a large temperature range (Albrecht et al., 1976). As depth and temperature increase the ratio of $(C_{15+}$ bitumens)/(kerogen) first increases (as in the California and Paris basins) but then reaches a maximum at about 6600 ft. and subsequently decreases (Figure 4.10). The increase reflects the generation of C_{15+} molecules from the kerogen. The decrease is caused by cracking of the C_{15+} molecules to give lower molecular weight products that are not included in the C_{15+} analytical range.

The cracking reaction may proceed in several different ways - the chain may be split in the middle or bond rupture may occur near one end.

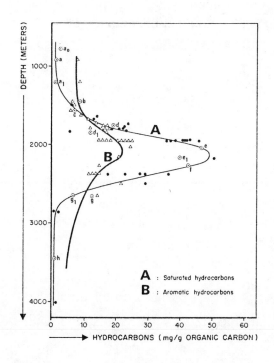

Figure 4.10:
Evolution of saturated (A:●) and aromatic (B:△) hydrocarbons with burial in the Lagbaba series. The amounts of saturated hydrocarbons from chloroform extraction as well as benzene-methanol extraction have been plotted. (Albrecht et al., 1976).

53

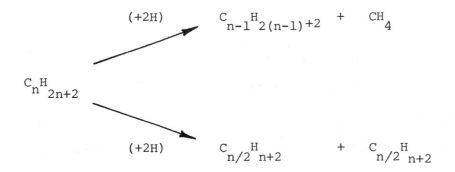

$$C_nH_{2n+2} \quad \xrightarrow{(+2H)} \quad C_{n-1}H_{2(n-1)+2} \quad + \quad CH_4$$

$$\xrightarrow{(+2H)} \quad C_{n/2}H_{n+2} \quad + \quad C_{n/2}H_{n+2}$$

Smith (1968) used a thermodynamic approach to establish that it takes much more thermal energy to break a C_4 chain in the middle than it does to break a C_6 chain, and this in turn takes more energy than C_8 median splitting. Above C_8 the differences are less pronounced. In general, it is more difficult to split off a short hydrocarbon group (C_1, C_2, C_3) than a longer one. This is consistent with the observation that gas accumulations are produced at greater depths than oil, where the higher temperatures provide the energy necessary to crack short chains. We may summarize these findings with an example. $C_{29}H_{60}$ occurs in leaves, etc. and at relatively low temperatures it will crack as follows:

$$C_{29}H_{60} \longrightarrow C_{14}H_{30} + C_{15}H_{30}$$

$$\downarrow {\scriptstyle +2H} \qquad\qquad C_{15+} \text{ extract}$$

$$C_{15}H_{32}$$

at intermediate temperatures,

$$C_{14}H_{30} \longrightarrow C_7H_{16} + C_7H_{14}$$

$$\downarrow {\scriptstyle +2H}$$

$$C_7H_{16} \qquad\qquad \text{products move out}$$

$$C_{15}H_{32} \longrightarrow C_7H_{16} + C_8H_{16} \qquad \text{of } C_{15+} \text{ fraction}$$

$$\downarrow {\scriptstyle +2H}$$

$$C_8H_{18}$$

at high temperatures,

$$C_8H_{18} \longrightarrow C_4H_{10} \quad + \quad C_4H_8$$
$$\downarrow +2H$$
$$C_4H_{10}$$

$$3\ C_7H_{16} \longrightarrow 3\ C_4H_{10} \quad + \quad 3\ C_3H_6$$
$$\downarrow +6H \qquad\qquad \text{wet gas}$$
$$3\ C_3H_8$$

The overall reaction is:

$$C_{29}H_{60} \quad + \quad 14H \longrightarrow 5C_4H_{10} \quad + \quad 3C_3H_8$$

Notice that the nature of the products at any stage depends on the temperature and that the products get lighter as temperature increases. Notice also that a source of hydrogen is required. This may be supplied by parallel reactions involving cyclization and aromatization of the kerogen (see Figure 4.7, for example).

Thermal processes are non-biogenic and operate to remove the characteristics of biogenically derived molecules. Paraffins generated by thermal cracking show no preference for odd or even chain lengths, and their increasing concentration eventually swamps the odd-even preference of the paraffins preserved directly from biogenic materials. Thus with increasing depth and maturity CPI values for normal paraffins decrease toward 1.0. Also, optically active compounds are degraded to give products showing no optical activity and the optical activity of rock extracts decreases as the active compounds are diluted and degraded. Many biologically derived molecules have complex and charac-teristic structures such as those found in the steranes, isoprenoids, porphyrins, etc. These compounds are eventually cracked to smaller molecules which lack the characteristics of their parents. The change

in the nature of the bitumens is widely used for indicating levels of

organic maturity.

Supporting evidence for the role of temperature in petroleum

generation comes from the study of sedimentary rocks intruded by igneous

dikes. If the intruded sedimentary sequence contains a shale, it pro-

vides an excellent opportunity for studying the effects of temperature

on rocks of the same age and type at the same pressure. Because of the

steep thermal gradient away from the dike the organic matter in the

rocks close to the intrusion have been exposed to high temperatures

while the organic matter farther away has been heated only a little

above the normal temperature of the country rock. Immediately adjacent

to the intrusion the organic matter is destroyed and only graphitic

material remains. Further out the extractable material is present in

the greatest quantities but then decreases smoothly with distance from

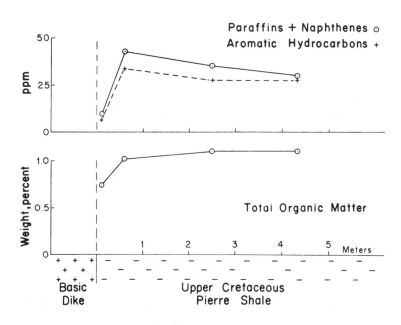

Figure 4.11:
Organic matter in the Upper Cretaceous Pierre Shale near an
igneous dike (Degens, 1965).

the contact, eventually reaching values characteristic of the host shale
(Figure 4.11). In this situation the role of temperature is unambiguous.

Further information on the effect of temperature on organic matter
is provided by laboratory heating experiments ("pyrolysis") which can
duplicate many of the features observed for subsurface samples. In
order to produce appreciable reaction in a reasonable length of time,
the temperatures used must be higher than those encountered in the
subsurface.

The effects of the generation process are summarized in Figure 4.12
which shows that biologically derived molecules form an important part
of the bitumens in shallow samples but become quantitatively less

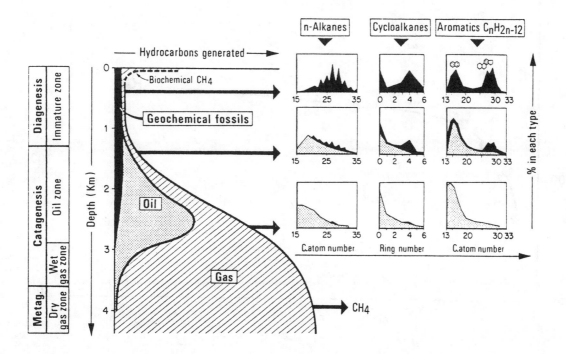

Figure 4.12: General scheme of hydrocarbon formation as a function
of burial of the source rock. The evolution of hydrocarbon composition is shown in insets for three different depths. The depth
scale is only approximate and will depend on geothermal gradient.
The values given correspond to an average for Mesozoic and
Paleozoic source rocks. Actual depths will also depend on organic
matter type and burial history (Modified after Tissot et al., 1974).

important in deeper samples. The bitumens start with strong biogenic
characteristics and show marked odd-carbon preference in the normal
paraffins, high abundances of 4- and 5-ring naphthenes and bimodal
distribution of aromatics. Generation produces a more mature extract
with a CPI closer to one and a distribution of multiringed naphthenes
and aromatics which more closely approximates thermodynamic equilibrium.
In the deepest samples the hydrocarbons become steadily lighter, giving
first wet gas and ultimately dry gas.

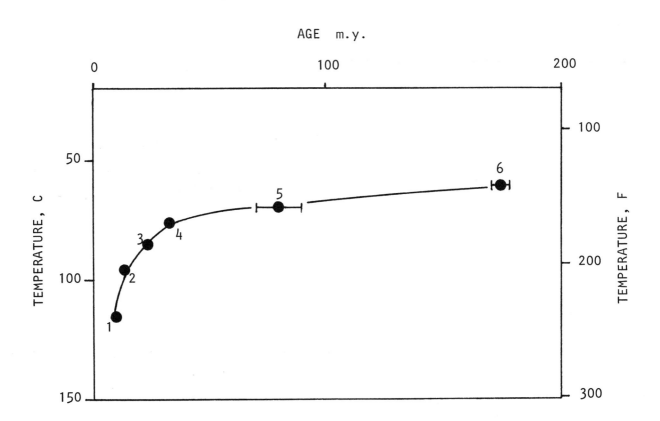

Figure 4.13: Temperature required for the onset of petroleum
generation in basins of different ages. Note that depth of
generation is much more sensitive to depth of burial for Tertiary
sequences. 1. Los Angeles basin (Philippi, 1965): 2. Louisiana
(LaPlante, 1974): 3. Louisiana (LaPlante, 1974): 4. Louisiana
(LaPlante, 1974): 5. Douala, Cameroons (Albrecht et al., 1976):
6. Paris basin, France (Tissot et al., 1971). Redrawn from
Schlanger and Combs (1975).

<u>Time</u>

Time has a role in the diagenesis of organic matter, but it is
secondary to that of temperature. Doubling the available time has the
same effect as increasing the temperature by approximately 10°C, so that

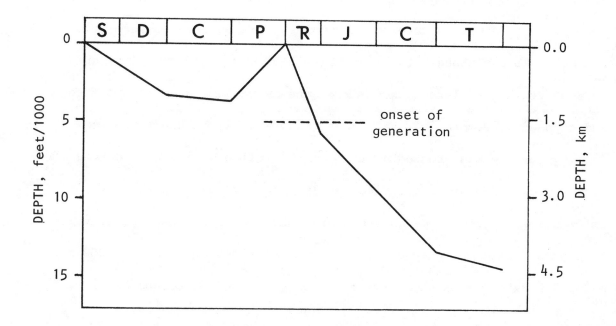

Figure 4.14: Burial depth as a function of geologic time for source
rocks of the Hassi Massaoud field, Algeria, showing that the
Silurian rocks were not buried deeply enough to generate until the
Late Triassic. (Redrawn from Poulet and Roucache, 1969).

the amounts of products increases linearly with increasing time but
exponentially for increasing temperature. In young sediments the onset
of generation will occur at greater depths (i.e. higher temperatures)
than it will in older sediments. Figure 4.8 shows   that for 12 m.y. old
Miocene sediments generation started at about 120°C while in the 175 m.y.
old Jurassic sediments of the Paris basin it started at about 60°C.
These data and others can be used to define a curve relating age to the
temperature required for petroleum generation (Figure 4.13).

It is not always easy to deduce the depth of generation because paleogeothermal gradients may have been very different from present values, and the sediments may have spent a considerable amount of time at higher or lower temperatures than the present ones. This is well illustrated by relationships for the Hassi Massaoud field, Algeria, where the crude oil is reservoired against a Triassic unconformity. The source rocks are thought to be Silurian (430 m.y. old) but after burial to 3000 ft. in the late Carboniferous they were subsequently uplifted and exposed by removal of overburden (Figure 4.14). When the unconformity surface was created at this time no oil was present, because the source rocks had not been buried deeply enough to generate, and so no oil was lost. Generation only occurred when the rocks were later buried to depths greater than 5000 ft. in the late Triassic. Thus over 200 m.y. elapsed between deposition of the source rocks and petroleum generation.

REFERENCES

Albrecht, P., Vandenbroucke, M. and Mandengue, M., 1976. Geochemical studies on the organic matter from the Douala Basin (Cameroon) - I. Evolution of the extractable organic matter and the formation of petroleum: Geochim. Cosmochim. Acta, v. 40, p. 791-799.

Blom, L., 1961. Thesis in D. W. van Krevelen, ed. "Coal: typology-chemistry-physics-constitution": New York, Elsevier p. 173.

Connan, J., 1974. Time-temperature relation in oil genesis: Bull. AAPG, v. 58, p. 2516-2521.

Correia, M., 1971. Diagenesis of sporopollenin: p. 569-620 in "Sporopollenin" (Ed. by Brooks, Grant, Muir, Van Gijzel and Shaw), Academic Press, London.

Degens, E. T., 1965. Geochemistry of sediments - a brief survey: Prentice-Hall, 342 p.

Dow, W. G., 1977. Kerogen studies and geological interpretations: J. Geochem. Explor., v. 7, p. 79-99.

Dunton, M. L. and Hunt, J. M., 1962. Distribution of low molecular weight hydrocarbons in recent and ancient sediments: Bull. AAPG, v. 46, p. 1-6.

Erdman, J. G., Marlett, E. M. and Hanson, W. E., 1958. The occurrence and distribution of low molecular weight aromatic hydrocarbons in recent and ancient carbonaceous sediments: paper presented at ACS Mtg. Chicago. Petrol. Div. Preprints c. C-39 - C-49.

Hunt, J. M., 1972. Distribution of carbon in crust of earth: Bull. AAPG, v. 56, p. 2273-2277.

LaPlante, R. E., 1974. Hydrocarbon generation in Gulf Coast Tertiary sediments: Bull. AAPG, v. 58, p. 1281-1289.

Philippi, G. T., 1965. On the depth, time and mechanism of petroleum generation: Geochim. Cosmochim. Acta, v. 22, p. 1021-1040.

Poulet, M. and Roucache, J., 1969. Geochemical study of the North Sahara reservoirs of Algeria (in French): Rev. Inst. Francaise Petr., May, p. 615-644.

Schlanger, S. O. and Combs, J., 1975. Hydrocarbon potential of marginal basins bounded by an island arc: Geology, July 1975, p. 397-400.

Smith, H. M., 1968. Qualitative and quantitative aspects of crude oil composition: Bur. Mines Bull. # 642, p. 1-136.

Tissot, B., Califet-Debyser, Y., Deroo, G. and Oudin, J. L., 1971. Origin and evolution of hydrocarbons in early Toarcian shales, Paris Basin, France: Bull. AAPG, v. 55, p. 2177-2193.

Tissot B., Durand, B., Espitalie, J. and Combaz, A., 1974. Influence of nature and diagenesis of organic matter in formation of petroleum: Bull. AAPG, v. 58, p. 499-506.

TOPIC 5

MIGRATION OF PETROLEUM FROM SOURCE ROCK TO RESERVOIR

INTRODUCTION

It is now generally accepted by geochemists that in sand-shale
sequences some of the petroleum generated in organic rich "source rocks"
can move to a porous and permeable reservoir rock where it may accumulate:

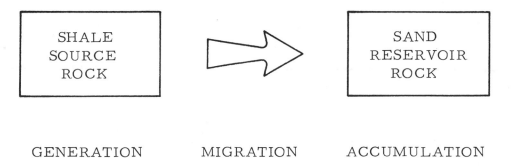

GENERATION   MIGRATION   ACCUMULATION

Thus migration links the source rock to the reservoir. An understanding
of the migration mechanism is important in many aspects of exploration
for the following reasons.

1. Correlation of reservoired crude oils to their source rocks is
based on the assumption that the composition of the source rock extract
should closely resemble that of the reservoired oil. This will be true
if the migration process causes only minor chemical fractionation. If
migration does produce considerable changes in the chemical composition
then many source rock-crude oil pairs may be missed by current techniques.

2. Many traps, particularly structural ones, develop later than
the onset of generation. Some knowledge of the time of migration is
important in establishing whether the trap was available to accumulate

the migrating petroleum.

3. The mechanism of migration is one of the factors controlling the distance of migration. This in turn sets limits on the maximum possible distance between the source rock and the trap.

4. Source rocks contain the material left after migration has occurred. A residue can only be interpreted if the composition of the migrated material is known and this depends on the migration mechanism.

5. The mineralogy of the source rock could have an important influence if internally generated water is necessary for migration. The presence of montmorillonite (which releases water when it is converted to illite) may be significant.

In spite of its obvious importance in the process of petroleum accumulation the migration of petroleum from source rock to reservoir is not well understood. Many migration mechanisms have been proposed (Cordell, 1972) and although all explain some of the observed features none seems generally applicable. Problems arise because hydrocarbons move in an aqueous medium yet have very low solubilities in water, and because movement through the very small pores of a fine-grained source rock puts severe limitations on the form in which the hydrocarbons can travel. Although the actual mechanisms of migration are not known some of the more reasonable suggestions are included in the following review.

MIGRATION MECHANISMS

True solution

Hydrocarbons have a very low solubility in water (Figure 5.1), but even a low solubility may be enough to move large amounts of petroleum if sufficient quantities of water are available. In general aromatics

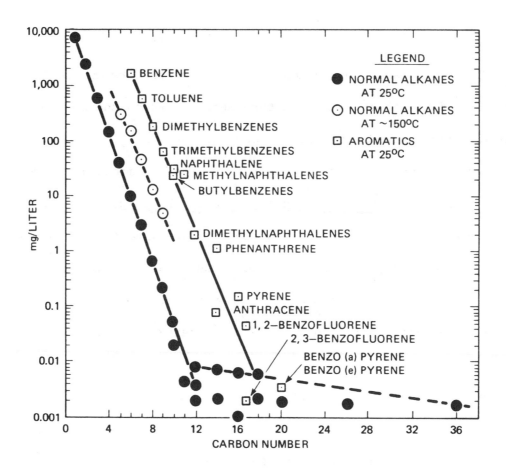

Figure 5.1:
Solubilities of normal alkane and aromatic hydrocarbons in water.
Summary from McAuliffe (1978).

are most soluble, paraffins least soluble and naphthenes intermediate,

with solubilities decreasing as the size of the molecule increases.

Most of the published hydrocarbon solubility data has been obtained in

fresh water at 25°C and one atmosphere pressure but solubilities de-

crease with increasing salinity and increase with increasing pressure,

particularly at high temperatures (Figure 5.2). Unfortunately only

limited data for solubilities under realistic subsurface condition of

salinity, temperature and pressure are available (Price, 1976).

Exsolution of hydrocarbons from the migrating aqueous solution

poses a major problem - why are the dissolved hydrocarbons not swept

through the trap with the water? Among the explanations suggested are
those involving changes in salinity, temperature, pressure and carrier
bed pore size.   It seems that the major discontinuity along the migration
path occurs at the sand-shale contact.   Here there are abupt changes in

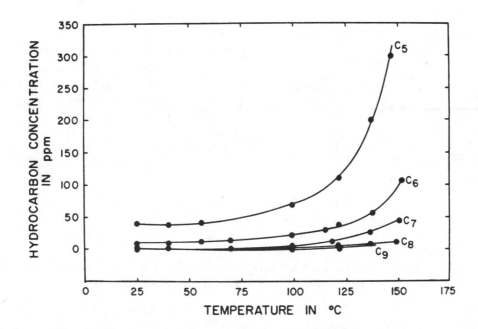

Figure 5.2: Solubilities of pentane (C_5), hexane (C_6),
heptane (C_7), octane (C_8), and nonane (C_9) in water as
function of temperature at systems' pressure. (Price, 1976).

water salinity and composition (Schmidt, 1973) so that pore water
saturated with dissolved hydrocarbons will be supersaturated in the more
saline waters of the adjacent sands (Figure 5.3). The exsolved droplets
of hydrocarbons could then be carried by moving waters through the sands
until they accumulate by buoyancy in a structural trap or are collected
at a permeability barrier in a stratigraphic trap.

Enhanced solubility

Several mechanisms have been suggested which involve the movement
of species more soluble than simple hydrocarbons. In these models

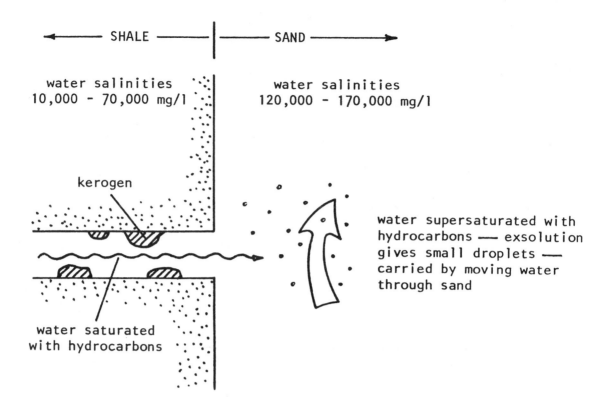

Figure 5.3: Exolution of dissolved hydrocarbons at a sand-shale contact. Salinity data from Schmidt (1973).

moving water is still required but smaller amounts can transport enough material to form a commerical accumulation. The problems of exsolution remain, but major effects can be expected at the sand-shale interfaces.

Accommodation: Accommodation is a term used for the suspension of minute colloidal particles (0.1 to 0.001µ diameter) in water. Peake and Hodgson (1967) pointed out that n-alkanes can be accommodated in distilled water as colloidal particles in amounts much greater than their true solubility. The lighter liquid n-alkanes are accommodated to a larger degree than the heavier solid n-alkanes. However, these same lighter n-alkanes (C_{12} and lower) are more readily disaccommodated by settling or filtering, hence a net preferential accommodation of C_{16}-C_{20} n-alkanes is observed. Mixtures of two or more n-alkanes increase the level of accommodation over the accommodation of the individual n-alkanes. The n-alkanes are accommodated in direct proportion to their abundance, and any odd predominance present in the source will be preserved in the water medium, and hence in any subsequently-formed oil accumulation.

Petroleum precursors: It has been suggested that it is not the petroleum hydrocarbons themselves that migrate but some more water-soluble precursor. Groups with NSO-containing functional groups, such as acids and alcohols, have high solubilities in water (Table 5.1) but the timing of the formation of hydrocarbons by loss of the functional groups would be critical.

Table 5.1:
Solubilities of paraffins, acids and alcohols in water (ppm).

	$R=C_5H_{11}$	$R=C_6H_{13}$	$R=C_7H_{15}$
R - H	39	9.5	3
R - COOH	10,700	2,200	-
R - OH	26,000	5,600	1,800

Micelles: Polar organic molecules may form small colloidal aggregates called "micelles" (Figure 5.4) with the water-compatible polar ends oriented out into the water. The hydrocarbon-like volume in the center can incorporate other organic molecules enhancing their apparent solubility. For this reason micelle-forming compounds are often referred to as "solubilizers" (Baker, 1960; Cordell, 1973).

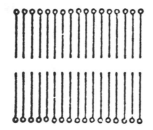

Figure 5.4:
Two configurations of
a micelle as proposed
by Baker (1960).

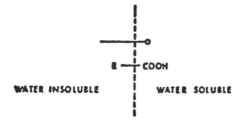

R —⊢— COOH

WATER INSOLUBLE   WATER SOLUBLE

Separate oil phase

If hydrocarbons are generated in quantities sufficient to saturate

the water as well as the adsorptive capacities of the shales and organic

matter, discrete droplets of crude oil may form in the pore spaces. As

water is expelled from the shale oil droplets will be carried out into

the coarser-grained carrier beds. This mechanism may operate if enough

moving water is available and if the diameter of the oil drop remains

less than the diameter of the pore. When the droplet encounters a con-

striction narrower than the drop diameter, deformation must occur if it

is to pass through (Figure 5.5). Deformation involves an increase in

surface area and hence an increase in surface energy. The only forces

operating on the droplet are bouyancy, which acts vertically upwards,

and the hydrodynamic gradient. Hill (quoted by Levorsen, 1954) has

shown that the hydrodynamic gradient is many orders of magnitude too

small to force the droplet through the constriction, while the bouyancy

forces have only a small component in the direction of movement through

near-horizontal pores. It has usually been concluded that droplets cannot negotiate the pore system of a shale. However, this simple model may have to be modified if the oil droplet traps and isolates a volume of water within a pore (as illustrated in Figure 5.6) because the temperature rise which accompanies increasing depth of burial (and causes generation) results in high overpressure in the isolated volume, and this could force the droplet through the constriction. A pressure differential of about 1 atm. is needed, and this will be generated aquathermally by a depth increase of about 10 ft. (Barker, 1972). For this mechanism to be important the droplet must effectively block the constriction and the volume of isolated water must be large compared to the droplet. A separate repressuring is needed at each constriction. These conditions are probably too restrictive for the mechanism to be generally important. However, the theoretical treatment is independent of scale and this mechanism may become important if the edges of a thick shale unit dewater preferentially so that average pore size is reduced near the edges. The center can then overpressure and force the release of both water and hydrocarbon droplets.

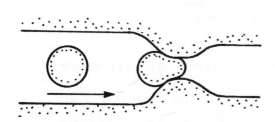

Figure 5.5: Deformation of an oil drop as it enters a constriction.

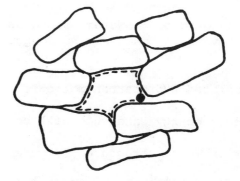

Figure 5.6: Schematic diagram showing how a droplet of oil may isolate a volume of water.

Pore-center Network

The development of a separate oil phase depends on the relative amounts of hydrocarbons and water. Estimation of the effective water volume is not straightforward because in the pores of the source rock the clay surfaces interact strongly with the adjacent water through hydrogen bonds. These bonds between the oxygen of the clay SiO_4 tetrahedra and the hydrogen of the water molecules orient the water close to the surfaces and "structure" it (Drost-Hansen, 1969) (Figure 5.7). The degree of structuring decreases with distance from the clay surface.

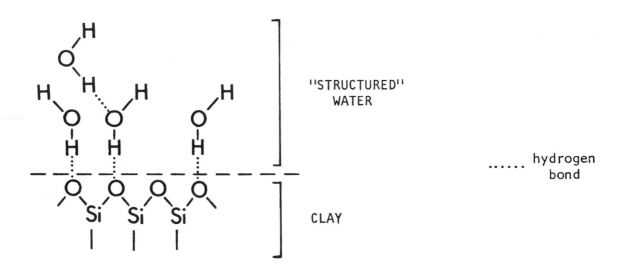

Figure 5.7: Water molecules near clay surfaces are held by hydrogen bonds and no longer behave like bulk liquid water.

An isolated hydrocarbon molecule in a pore will assume a configuration of least energy, and this will be in the center of the pore where the water is least structured (Figure 5.8A). If a second hydrocarbon molecule enters the system it may lie alongside the first (Figure 5.8B) or end-to-end with it (Figure 5.8C). Additional hydrocarbon molecules should eventually lead to the situation represented in Figure 5.8D. Hydrocarbon molecules do not suddenly appear in the pores, but are

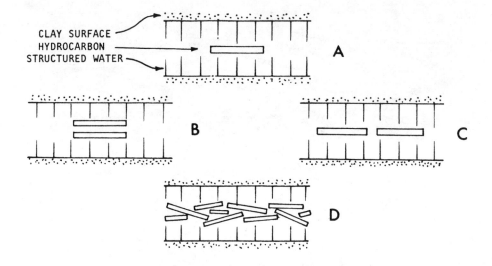

Figure 5.8

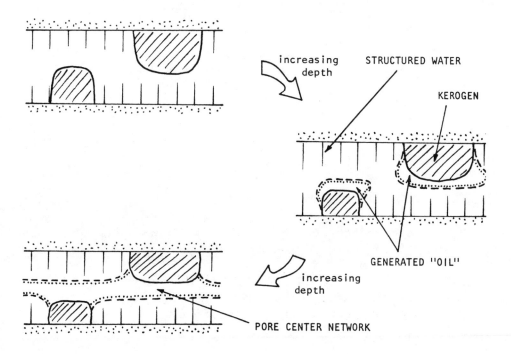

Figure 5.9: Development of pore-center network.

generated from the kerogen. A possible sequence of events is shown in Figure 5.9. The generated hydrocarbons are initially adsorbed onto the kerogen but as generation continues they become relatively more abundant and build out into the water. Preferential replacement of the less

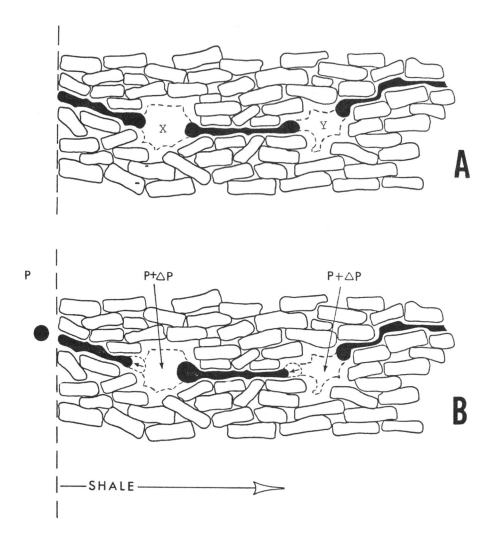

Figure 5.10: Possible distribution of oil as a pore center network
in a shale adjacent to a sand.
A. Distribution of oil and water at 10,000 ft. and 85°C.
B. Burial of the shale to 14,000 and 115°C. causes thermal
expansion of the water in volumes X and Y, and leads to
the expulsion of part of the hydrocarbons.

structured water causes lateral development until finally they merge

with hydrocarbons generated by other pieces of kerogen to form a pore-

center network (Barker, 1973).

The networks developed in this way will probably be discontinuous

because the irregular packing of clay platelets causes large variations

in pore diameter throughout the shale, and in the larger cavities there will be insufficient hydrocarbons to continue the network. This situation is shown schematically in Figure 5.10 where hydrocarbon networks have developed in the pores but do not extend across the larger voids.

Even after the formation of pore center networks some pressure differential or other force is needed to produce a net movement of hydrocarbons out of the rock. Again we must recognize that the system is not isothermal, and that petroleum generation is generally caused by rising temperature. The water and other fluids in the pores will expand as the temperature rises and create an internal pressure in the shales. Since the water is structured it will exhibit non-Newtonian behavior and will resist movement, while the hydrocarbons will be squeezed out. This process operates even when the network is not continuous. Consider, for example, Figure 5.10B. If the temperature rises from $85^{\circ}C$ to $115^{\circ}C$ (equivalent to increasing the depth of burial from 10,000 ft. to 14,000 ft. in the Gulf Coast), the water density changes from 1.0165 cm^2/g to 1.0317 cm^2/g corresponding to a 1.5 percent expansion at constant pressure. If the water in volume X expands by 1.5 percent it will force an equivalent volume of oil through the narrower pore structure into the next large opening Y, so that this receives an influx of oil equivalent to 1.5 percent of the volume of X. However, the water in Y is also subjected to increasing temperature and it expands by 1.5 percent of the volume of Y. This increase in volume is added to that produced in volume X so that the volume of oil expelled to the adjacent sand is approximately 3% of the volume of Y if cavities X and Y are of equal volume. It appears that the volume of oil squeezed from the rock by the thermal expansion of water is equal to 1.5 percent of the total pore space when the temperature rises from 85° to $115^{\circ}C$, and the pressure

73

from 5000 psi (10,000 ft.) to 7000 psi (14,000 ft.) even when the pore network is discontinuous. If this mechanism operates with 100 percent efficiency, it leads to unrealistically high values for the percent of the bitumens that migrates. For an average shale with 1 percent kerogen this will occupy about 2.5 percent of the rock volume and generate bitumens that occupy roughly 0.25 percent of the rock. For a shale with 10 percent porosity the increase in volume of the water on heating (again using 85° to 115°C and 10,000 to 14,000 ft.) will be 1.5 percent, or 0.15 percent of the volume of the rock. If only bitumen moves, 0.15/0.25 of it will be expelled, which is 60 percent of the available bitumen. Previous studies of migration efficiency showed that only 1-10 percent of the bitumens migrated. This leads to the conclusion that the pressuring mechanism discussed above only needs to be about ten percent efficient to drive out enough bitumens to form commercial deposits.

It should be noted that the theory for aquathermal pressuring was developed for liquid-filled systems (Barker, 1972). If the systems (such as volume X, Figure 5.10) contain more than about 5 percent gas they will not overpressure, and no net pressure differential will be developed to drive migration. Although simple thermal expansion of a gas-containing isolated volume does not overpressure, the continued generation of gas from the thermal maturation of kerogen may develop sufficient pressure (Momper, 1978).

Miscellaneous mechanisms

A wide variety of other mechanisms have been discussed in the literature. Some of these may be important under some situations. For example, the role of high pressure gas in transporting other hydrocarbons presumably will be most important deep where gas generation is most pronounced (Sokolov and Mironov, 1962), but the role of gas at shallower depths where it may be generated from woody-type kerogens is unknown. Diffusion of hydrocarbons, which permits migration without a moving water phase, is generally considered to be incapable of moving hydrocarbons over large enough distances. However, it could have a role in moving materials short distances to fractures or sandy stringers where some other mechanism may take over. Continuous networks of kerogen distributed through the source rock have been postulated as one way of permitting transfer of hydrocarbons to the adjacent reservoir rocks (McAuliffe, 1978), but it is difficult to see how a network would be established initially and how it would survive rearrangement of grains during compaction and diagenesis since a single break in the network would stop the whole migration process.

	Straight Chains	Branched Chains	Cyclic Compounds
C_5	38.5 (Pentane)	47.8 (iPentane) 33.2 (DMPropane)	156 (cPentane)
C_6	9.5 (Hexane)	13.8 (2MPentane) 12.8 (3MPentane) 18.4 (2,2DMButane)	55 (cHexane) 42.6 (McPentane)
C_7	3 (Heptane)	4.1 (2,4DMPentane)	29.6 (cHexane) 14.0 (McHexane)

i, iso-; c, cyclo-; M,methyl; DM, dimethyl

Table 5.2: Solubilities in water (ppm) of some saturated hydrocarbons with five, six or seven carbon atoms (Data from McAuliffe, 1963, 1966).

SIGNIFICANCE OF CORRELATION STUDIES

In many basins crude oils can be related to the specific rock units which generated them because there is a close similarity between the composition of the bitumens in the source rock and the composition of the crude oil. This implies that, at least in these cases, migration causes only minor changes in chemical composition. This, in turn, puts severe constraints on the mechanism of migration.

In a detailed study of crude oils and possible source rocks in the Williston basin Williams (1974) identified three types of crude oils which could be correlated to Winnipeg, Bakken and Tyler source rocks respectively. Correlation was based on C_4-C_7 hydrocarbons, C_{15+} hydrocarbons, carbon isotope values, optical rotation and Correlation Index plots. Figure 5.11 shows how closely the compositions of the source rock extracts resemble those of the crude oils when plotted on a straight chain-branched chain-saturated ring compound ternary diagram. The solubilities of hydrocarbons in these three compound types vary considerably (Table 5.2), and any migration mechanism involving a solubility mechanism should lead to marked differences in the chemical composition of the oil relative to the source rock extract. Since this is not observed it strongly suggests that a mechanism involving solution in water is not appropriate in the Williston basin. Data for the C_{15+} range of straight chain hydrocarbons also supports this conclusion because the source rock extract and the oils have very similar abundance patterns (Williams, 1974).

However, it is dangerous to generalize from a single example and the Williston basin may not serve as a reliable guide for other areas. In many cases source rock extracts do not have the same composition as the oils, though they are often sufficiently alike that correlation is possible (Barker, 1975). In general, crude oils are enriched in saturate hydrocarbons relative to the aromatics and usually have less NSO compounds than the source rock extracts. Data for the Parentis Basin are presented in Figure 5.12 and show that relative to the source rock extract the oils are enriched in compounds in the sequence saturates > aromatics > NSO's. Interestingly this sequence is exactly the opposite of the aqueous solubilities - the oils are enriched in the least soluble components.

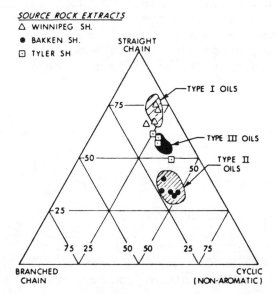

Figure 5.11:
Distribution of straight chain, branched and cyclic paraffins in the C_4-C_7 fraction of Williston basin crude oils and source rock extracts (Williams, 1974).

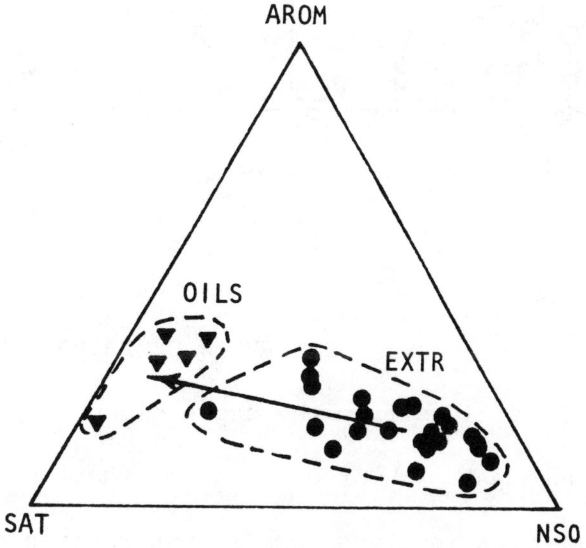

Figure 5.12:
Distribution of saturate, aromatic and NSO compounds in Jurassic crude oils and source rock extracts from the Parentis basin, France (Deroo, 1976).

Although it might be argued that the more soluble compounds are less likely to accumulate and continue to move with the transporting aqueous solution, it seems much more likely that adsorption in the source rock is playing a major role (Tissot and Pelet, 1971). Normal paraffins are adsorbed less strongly than aromatics or NSO's by both organic matter and mineral surfaces and they will move out of the rock preferentially. An example of the interaction of an oxygen-containing compound (an alcohol) with a clay surface is shown in Figure 5.13. Moving hydrocarbons through a rock is, in some ways, similar to moving a mixture through a chromatographic column and separations can be expected due to the different relative rates of movement.

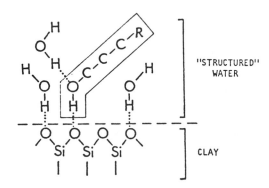

Figure 5.13: Interaction of the -OH group of the R-C-C-C-OH alcohol with a clay surface. Hydrogen bonding is involved in the same way as it is for the -OH groups in water molecules.

OVERVIEW OF PRIMARY MIGRATION

It follows from the above discussion of migration mechanisms that the movement of hydrocarbons, or their precursors, out of a fine-grained source rock involves complex interaction between petroleum, rock matrix and water. In Figure 5.14 an attempt has been made to summarize the important variations in these parameters as depth and temperature increase.

Water from compaction of fine-grained sediments is abundant at shallow depths but decreases with depth and is probably minor below 10,000 ft. Water can also be produced from smectite (montmorillonite), first by release of interlayer water and, slightly deeper, by conversion to illite through reaction with potassium ions. Only a generalized water-loss curve is shown in Figure 5.14. Not all source rocks contain abundant smectite and this deep source of water may not be available.

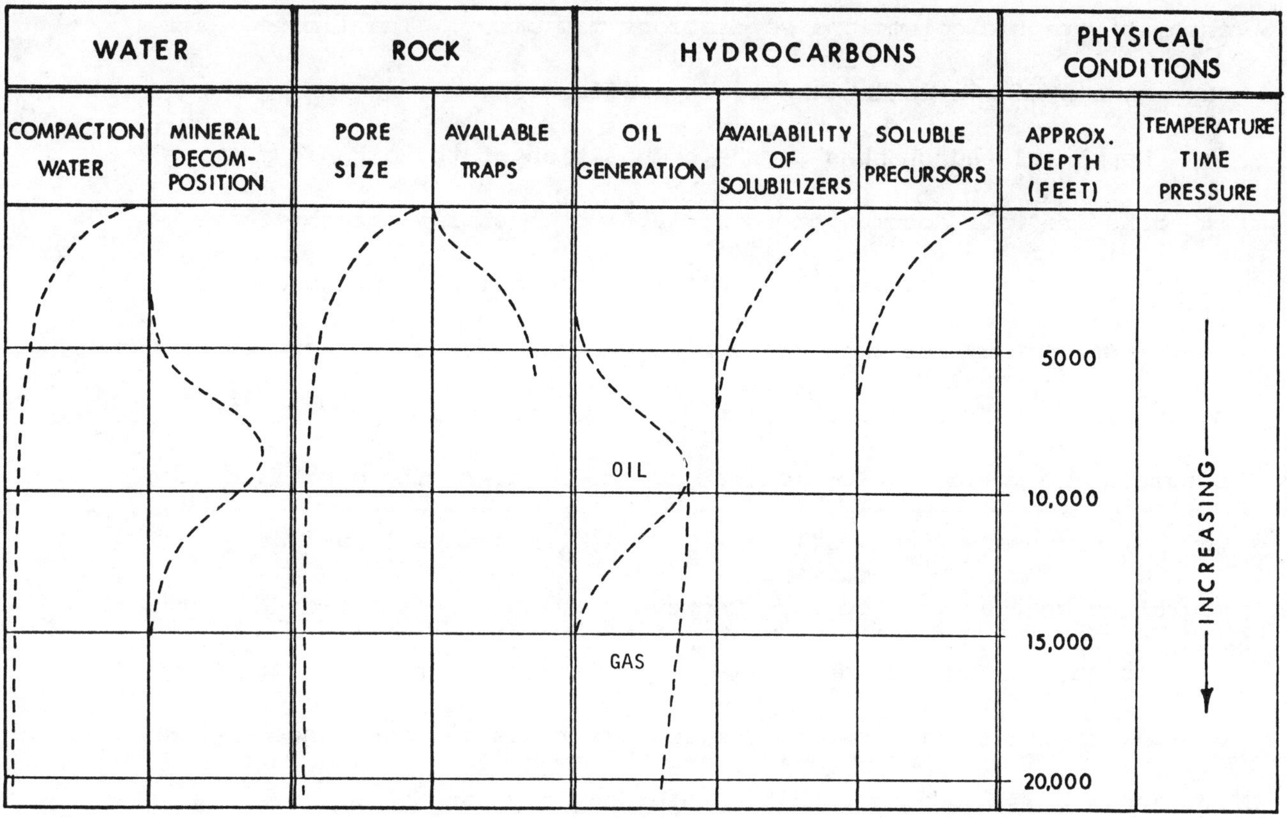

WATER		ROCK		HYDROCARBONS			PHYSICAL CONDITIONS	
COMPACTION WATER	MINERAL DECOMPOSITION	PORE SIZE	AVAILABLE TRAPS	OIL GENERATION	AVAILABILITY OF SOLUBILIZERS	SOLUBLE PRECURSORS	APPROX. DEPTH (FEET)	TEMPERATURE TIME PRESSURE

OIL

GAS

5000

10,000

15,000

20,000

INCREASING

Figure 5.14

The question of whether the extra water helps, hinders, or has no role in migration is unanswered, although it is certainly not essential for migration because many areas with no smectite have abundant petroleum.

The rock is the matrix through which migrating materials move. The pore structure of the rock, particularly the average pore size, puts constraints on the form of the migrating material. Much less is known about the pore structure of shales than sands but average pore size is reduced as compaction proceeds and diagenetic processes such as cementation can also reduce pore volume. The chemical interaction of mineral surfaces (and kerogen) with migrating organic matter is not well documented for subsurface conditions even though the movement of organics through an interacting rock is comparable to moving them through a

79

chromatrographic column and separations may occur. The limited data suggest a preferential movement of paraffins.

Depth-related changes in the composition of the organic materials in rocks are well documented (Dow, 1977). At shallow depths there may be some bacterial methane along with the biologically produced lipids (bitumens) but hydrocarbons generated thermally from the kerogen do not become important until depth of burial exceeds about 6000 ft. The bitumens are initally NSO-rich but become lighter with depth and give way to condensate and finally gas. Thus, the material available for migration changes with depth. Compounds which can form micelles or move as petroleum precursors generally have functional groups, like -COOH or -OH, but these are lost as temperature rises and the concentration of such compounds, and any possible role they may have in petroleum migration (such as micelle formation), decreases with depth.

All these changes usually occur while temperature, pressure and time (age) are increasing and the physical properties of organic materials such as solubilities, viscosities, densities, etc. are being altered. We can discuss the proposed migration mechanisms outlined above against this background information for the rock-organic matter-water system.

In the first few thousand feet water from compaction is abundant and pore size is relatively large, but temperatures are too low for significant generation of petroleum and the only hydrocarbons present are those preserved from once-living systems. These contain no C_2-C_{10} hydrocarbons but many of the larger molecules have functional groups which enhance their solubilities so that they can move with the water. Trapping efficiency seems to limit accumulation. Kidwell and Hunt (1958) found that early accumulations in the Pedernales area of the

Orinico delta were much enriched in aromatics relative to paraffins.
This is the expected relationship for water-transported organics since
the aromatics in general have high solubilities. Bacterially-generated
methane also may be present in shallow sediments and some commerical gas
accumulations have been formed from this source. Presumably the high
solubility of the methane enhances migration but movement of a separate
gas phase by buoyancy is also possible.

As depth of burial exceeds five or six thousand feet thermal
generation of hydrocarbons from kerogen becomes important and traps are
common, but the available water is limited, average pore size is diminished
and the concentration of micelle formers and soluble hydrocarbon precursors has decreased. The commerical accumulations which form at these
depths tend to be heavy with high nitrogen, sulfur and oxygen contents
since thermal diagenesis of kerogen has not yet produced appreciable
quantities of the low molecular weight hydrocarbons. These are formed
in abundance as depth and temperature continue to increase and by about
10,000 ft. higher ^{O}API gravity crudes are common. The water available
from compaction of sediments at these depths is minor, although minera-
logic changes can provide large quantities. This "second squirt" will
be available only in certain locations but, as noted above, petroleum
occurs in areas where this source of water has not been available. At
these depths simple molecules with functional groups (such as acids and
alcohols) will no longer be present to form micelles or move as precursors because reactions such as decarboxylation remove the function
groups which confer water solubility. Micelles are unlikely to be
important anyway because the average pore size is now less than the
diameter of a micelle which would restrict their movement to the larger

pores and fractures leaving large amounts of potentially productive source rock undrained. Lack of water and progressing generation may lead to the development of a separate crude oil phase in the pores. Pore center network formation may also develop at this stage allowing transfer of hydrocarbons out of the shale.

In the deepest part of basins crude oil gives way to condensate and ultimately gas (Topic 6) so that the material to be moved changes greatly in physical properties. It is here that mechanisms involving gas solution seem most plausible. Increasing temperatures also enhance the solubilities of hydrocarbons.

It should be obvious from this brief discussion that the nature of the material to be moved varies considerably, and the physical nature of the medium through which the migration occurs also shows marked changes with depth. The idea that there is a single mechanism for the migration of petroleum out of a source rock is almost certainly wrong and several different mechanisms may operate at different times and at various stages of burial and generation. The various mechanisms proposed for petroleum migration should not be regarded as alternative suggestions, but rather as complementary ones.

EFFICIENCY OF MIGRATION

The overall processes of migration and trapping seem to be very inefficient. The estimates of total extractable hydrocarbons are more than two orders of magnitude greater than the estimates of total reservoired petroleum, showing that less than one percent of the hydrocarbons in source rocks accumulates in reservoirs. Detailed mass balance studies in single basins have all produced values of a few percent for the amount migrated (e.g. Hunt, 1977). The upper limit from these studies is about 10 percent. This suggests that the process of migration will not produce major changes in the composition of the source rock extract.

SECONDARY MIGRATION

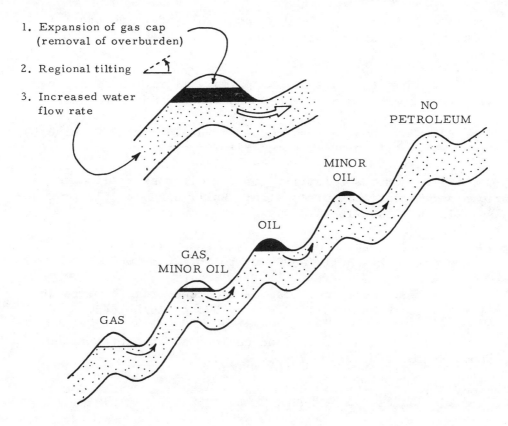

Figure 5.15: Schematic representation of the possible results of differential entrapment.

The remobilization of crude oils after they are initally reservoired is called "secondary migration". It is most frequently caused by regional tilting. Effects can range from a simple relocation of the petroleum in a new reservoir to a marked charge in chemical composition. Gussow (1954) has coined the term "differential entrapment" for the case where a full trap spills petroleum from the bottom into the next higher trap (Figure 5.15). This can lead to adjacent traps with gas, oil or varying mixtures of the two. The gas-filled reservoirs will be down dip from the oil-filled ones.

REFERENCES

Baker, E. G., 1960. A hypothesis concerning the accumulation of sediment hydrocarbons to form crude oil: Geochim. Cosmochim. Acta, v. 19, p. 309-317.

Barker, C., 1972. Aquathermal pressuring - role of temperature in development of abnormal-pressure zones: Bull. AAPG, v. 56, p. 2068-2071.

Barker C., 1973. Some thoughts on the primary migration of hydrocarbons: Paper presented at the AAPG Ann. Mtg., San Antonio.

Barker, C., 1975. Oil source rock correlation aids drilling site selection: World Oil, Oct. 1975, p. 121-126, 213.

Barker, C., 1977. Aqueous solubility of petroleum as applied to its origin and migration: Discussion: Bull. AAPG, v. 61, p. 2146-2149 (see also p. 2149-2156).

Cordell, R. J., 1972. Depths of oil origin and primary migration: a review and critique: Bull. AAPG, v. 56, p. 2029-2067.

Cordell, R. J., 1973. Colloidal soap as proposed primary migration medium for hydrocarbons: Bull. AAPG, v. 57, p. 1618-1643.

Dow, W. G., 1977. Kerogen studies and geological interpretations: J. Geochem. Explor., v. 7, p. 79-99.

Drost-Hansen, W., 1969. Structure of water near solid interfaces: Ind. Eng. Chem., v. 61, p. 10-47.

Gussow, W. C., 1954. Differential entrapment of oil and gas: a fundamental principle: Bull. AAPG, v. 38, p. 816-853.

Hunt, J. M., 1977. Ratio of petroleum to water during primary migration in western Canada basin: Bull. AAPG, v. 61, . 434-435.

Kidwell, A. L. and Hunt, J. M., 1958. Migration of oil in recent sediments of Pedernales, Venezuela: in "Habitat of Oil" (Ed. by L. G. Weeks), p. 790-817.

Levorsen, A. I., 1954. Geology of petroleum: Freeman, San Francisco. 703 p.

McAuliffe, C. D., 1963. Solubility in water of C_1-C_9 hydrocarbons: Nature v. 200, p. 1092-1093.

McAuliffe, C. D., 1966. Solubility in water of paraffin, cyclo-paraffin, olefin, acetylene, cyclo-olefin, and aromatic hydrocarbons: J. Phys. Chem., v. 70, p. 1267-1275.

McAuliffe, C. D., 1978. Role of solubility in migration of petroleum from source: p. Cl-C39 in "Physical and Chemical Constraints on Petroleum Migration" (Ed. by W. H. Roberts, III and R. J. Cordell). AAPG Continuing Education Course Note Series #8.

Momper, J. A., 1978. Oil migration limitations suggested by geological and geochemical considerations: in "Physical and Chemical Constraints on Petroleum Migration" AAPG Continuing Education Course Note Series #8.

Peake, E. and Hodgson, G. S., 1967. Alkanes in aqueous systems. I. The accommodation of C_{12}-C_{36} n-alkanes in distilled water: J. Amer. Oil Chemists Soc., v. 44, p. 696-702.

Price, L. C., 1976. Aqueous solubility of petroleum as applied to its origin and primary migration: Bull. AAPG, v. 60, p. 213-244.

Roberts, W. H., III and Cordell, R. J., 1978. "Physical and Chemical Constraints on Petroleum Migration": AAPG Continuing Education Course Note Series #8.

Schmidt, G. W., 1973. Interstitial water composition and geochemistry of deep Gulf Coast shales and sandstones: Bull. AAPG, v. 57, p. 321-337.

Sokolov, V. A. and Mironov, S. I., 1962. On the primary migration of hydrocarbons and other oil components under the action of compressed gases: in The chemistry of oil and oil deposits: Acad. Sci. USSR, Inst. Geol. and Exploit. Min. Fuels, p. 38-91 (in Russian); 1964 Engl. trans. by Israel Progr. Sci. Trans., Jerusalem.

Tissot, B. and Pelet, R., 1971. Nouvelles donnees sur les mecanismes de genese et de migration du petrole simulation mathematique et aplication a la prospection: Proc. 8th World Pet. Congr., v. 2, p. 35-46.

Williams, J. A., 1974. Application of oil-correlation and source rock data to exploration in the Williston basin: Bull. AAPG, v. 58, p. 1243-1252.

TOPIC 6

MATURATION AND ALTERATION OF RESERVOIRED PETROLEUM

GROSS VARIATIONS IN COMPOSITION

The chemical composition of a crude oil is not fixed but changes in response to changing conditions in much the same way as kerogen composition changes (Topic 4). The evolution of a crude oil involves a process of continuous, irreversible change, usually called "maturation", which leads from a heavy, NSO-rich "immature" crude oil towards "mature" crudes which are lighter. This trend was recognized as early as 1915 when White found that in areas where coal and oil occur together the higher rank coals were associated with the oils of highest API gravity. Since rank was known to increase with temperature it followed that API gravity also increased with temperature. Barton (1934) documented this trend for Gulf Coast oils where the deepest (and oldest) reservoirs contain lighter and more paraffinic crudes. Similar trends have been observed in many other areas also (although they are not universal) (Figure 6.1).

In documenting the effects of increasing depth on maturation, data is accumulated for many different crude oils taken from reservoirs located over the entire depth range. There are inherent uncertainties in this approach because crude oils generated by different source rocks may have started with very different compositions, although statistical studies involving large numbers of crude oils will tend to smooth out these effects. An alternative approach to the study of the effect of increasing depth of burial on composition is to examine the changes produced in a single crude oil type. Data of this type have been reported by Koons et al. (1974) (Table 6.1) and Orr (1974) (Figure 6.2).

Table 6.1: Variation in composition of related Tuscaloosa oils with
depth (Koons, et al., 1974).

Field	Bottom Hole Temperature	$\frac{Cyclo\ C_5}{Cyclo\ C_6}$	Percent Paraffins (in sats.)	Percent Polar (in C_{15+})	δC^{13} (sats.)
Dexter	232°F	1.7	22	11	-28.3
Little Creek	238	2.5	27	4	-28.2
Mallslieu	242	2.2	25	8	-28.2
Smithdale	249	2.5	24	8	-27.9
West Bude	256	2.4	29	2	-27.9
Carthage Point	254	3.0	31	1	-27.3

A considerable amount of data is now available showing that crude oils change in composition in response to changing physical conditions in the reservoirs. The various factors which can cause these changes are best considered under two headings. First, "maturation": this term is used for all processes which occur as an inevitable consequence of increasing age and depth of buial. Secondly, "alteration": this is caused by an external influence (usually chemical or bacterial) which acts to change the composition of the crude oil.* All crude oils are subject to maturation but the effects of alteration may vary widely from one crude oil to another or from one part of a reservoir to another. In many cases the effects of alteration are absent.

MATURATION: CAUSES AND EFFECTS

Temperature

Temperature has a major role in maturation, and thermal cracking is the most important mechanism. Crude oils are not equilibrium mixtures and as temperature increases they readjust towards equilibrium at an increasing rate. This readjustment takes the form of a redistribution

---

*It should be noted that some authors use "thermal alteration" as a synonym for "maturation."

87

of hydrogen (Figure 6.3) ultimately giving methane and a solid carbon-rich residue in a process similar to that discussed for kerogen diagenesis. The sequence of compositions leading from immature crude to methane has been summarized by Dobryanski (1968) into the scheme shown in Figure 6.4. The vertical scale is one of increasing maturation downwards. The

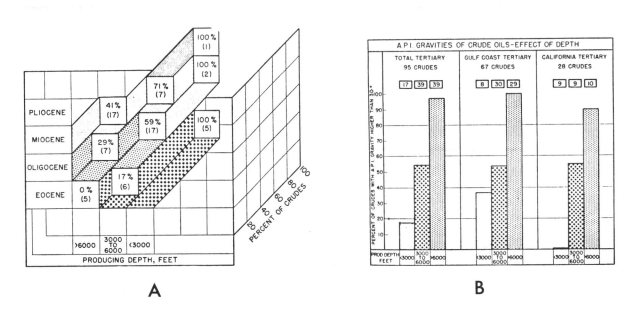

Figure 6.1: Effect of age and depth on low molecular weight fractions and API gravity of crude oils. A: Percent of crudes with 20% at 392°F, Gulf Coast Tertiary (figures in parentheses give number of crudes in group). B: Percent of crudes with API gravities >30° for various depth intervals. (McNab et al., 1952).

composition at Stage VIII, for example, is 13% natural gas, 57% paraffinic hydrocarbons, 2% monocyclic naphthenes, 2% polycyclic naphthenes, 6% monocyclic aromatic hydrocarbons and 20% secondary resinous substances. As maturation proceeds through to Stage X the amount of natural gas (methane) increases at the expense of the other compound types. However, methane is a very hydrogen-rich compound (H/C = 4) and there is insufficient hydrogen in crude oils to convert all the carbon to methane. The net effect is to leave a carbon-rich residue which

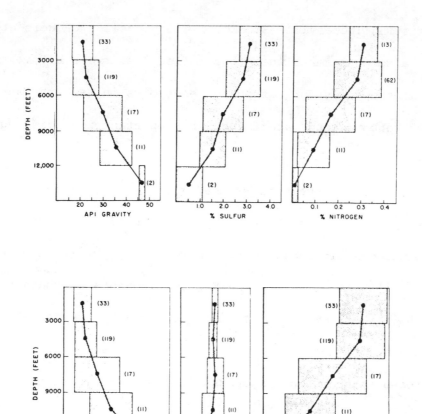

Figure 6.2: Depth trends in Big Horn basin Paleozoic oils
(U. S. Bur. Mines Data for 182 samples). Points indicate
mean for 3,000 ft. interval; bar width is one standard
deviation (Orr, 1974).

Dobryanski calls "highly carbonaceous" and "graphitic" compounds.

The products which get progessively richer in carbon eventually

precipitate out and are left in the reservoir. For example, the Thomas-

ville field produces sour, thermal gas from 20,000 feet in the Smackover

and the reservoir has been described by Parker (1973) as a "black vuggy

sandstone". He refers to the black material as "an infusible solid

asphaltic hydrocarbon of high fixed carbon ratio" and argues that the

"methane and black residue represent products from a thermal metamorphism of a pre-existing Thomasville-Smackover oil deposit". Rogers et al. (1974) have also considered what they called "reservoir bitumens" in some detail, especially for carbonate reservoirs in Western Canada. The precipitated carbonaceous material can be important in modifying reservoir porosity.

Oils occurring in the Leduc-Rimbey reef trend of Alberta are members of a single family but they have been matured to different extents in different geographic locations (Milner et al., 1977). Geochemical data for six oils are summarized in Figure 6.5 and show that as depth increases the higher temperatures have led to higher percentages of paraffins and naphthenes, decreased content of polar compounds and a slight decrease in aromatic compounds with four rings or less. The

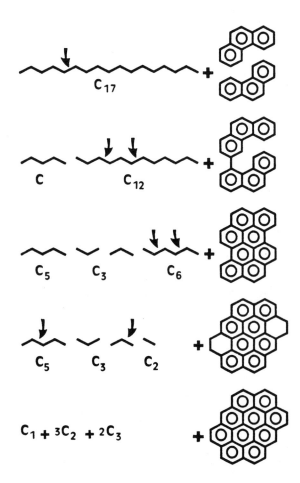

Figure 6.3: Schematic representation of the thermal maturation of a crude oil showing the parallel development of increasingly aromatic heavy molecules and progressively smaller paraffins (after Connan et al., 1975).

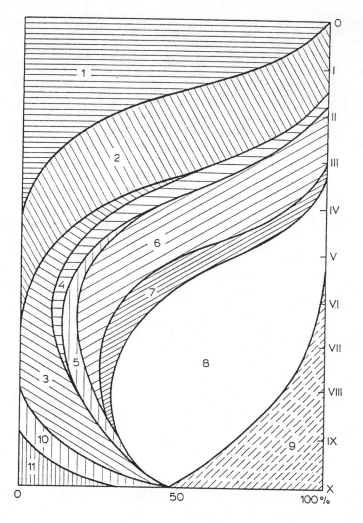

Figure 6.4: Schematic representation of the changing chemical composition of maturing crude oil. Diagram shows the group composition of petroleum as a funcation of degree of conversion. *1* deoxygenated initial petroleum substance; *2* primary resinous substances; *3* secondary resinous substances; *4* polycyclic aromatic hydrocarbons; *5* monocyclic aromatic hydrocarbons; *6* polycyclic naphthenes; *7* monocyclic naphthenes; *8* paraffinic hydrocarbons; *9* natural gas; *10* highly carbonaceous compounds; *11* graphitic compounds.

percentage of the C_{15+} fraction decreases as light ends are produced. (The Nestow oil shows evidence of compositional changes produced in part by bacterial degradation (see below)).

The trend from oil to gas with increasing temperature implies that there is some definite temperature above which oil will not be found. On the other hand it takes a certain amount of temperature for oil generation to begin so that oil is only available over a limited temperature range. Pusey (1974) has called this the "oil window". His figure (based on earlier work by Landes (1967) and reproduced here as Figure 6.6) shows how the depth of the oil window varies with geothermal gradient.

Note that in areas with high geothermal gradient oil can be generated shallower but the total oil reserves are likely to be less because the depth interval between generation and destruction is smaller. Pusey's representation is limited because it does not show the role of time. The temperature needed for the onset of petroleum generation is modified by the effects of geological age (as shown in Figure 4.13, page 58) and this effect has been included in extending the "oil window" concept as

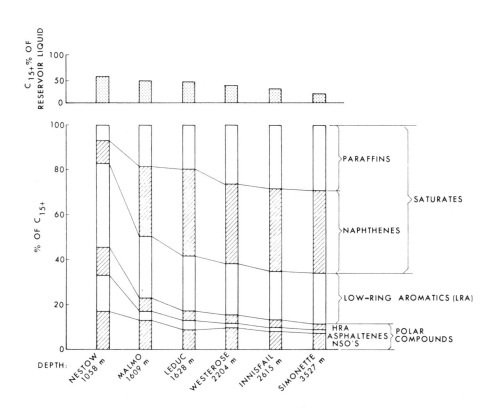

Figure 6.5: Liquid chromatographic separations obtained from C_{15+} fractions of six Leduc reef oils, Alberta. Upper histogram shows total C_{15+} content. Lower histogram shows weight percent of individual compositional classes within C_{15+} fraction: saturates, consisting of paraffins and naphthenes; low-ring aromatics (four rings or fewer); polar compounds, consisting of high-ring aromatics (more than four rings), asphaltenes and NSO's (molecules containing N, S, or O in addition to C and H). Note that the produced oil is only a partial product and heavier residues (such as those produced from the polar compounds) may be left in the reservoir. There is no mass balance for the compositional changes illustrated.

shown in Figure 6.7.

Oils get lighter with depth because larger molecules are broken down into smaller ones. This involves carbon-carbon bond cleavage. Carbon has two stable isotopes and so bonds can be either $^{12}C-^{12}C$ or $^{13}C-^{12}C$ ($^{13}C-^{13}C$ bonds are quantitatively insignificant) but it takes slightly more energy to break the bonds containing ^{13}C atoms and they

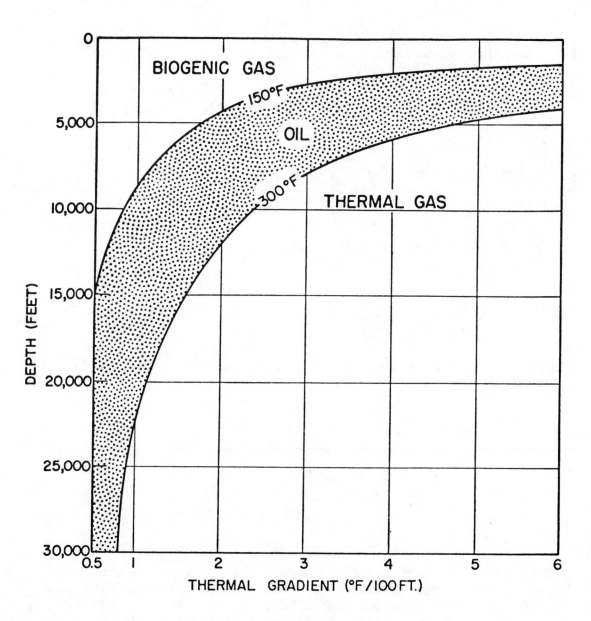

Figure 6.6

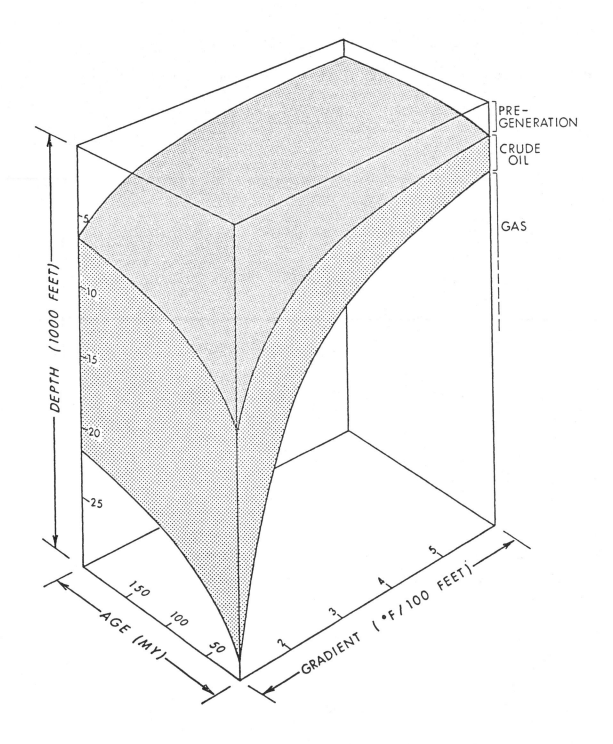

Figure 6.7: The generation and destruction of crude oil related to depth, age and geothermal gradient. The top surface corresponds to oil generation and the bottom surface to destruction. Neither should be taken as sharp boundaries but rather as gradational changes.

94

rupture less frequently. Because of this the lower molecular weight hydrocarbons which are broken off are enriched in ^{12}C (i.e. are iso-topically ligher), while the higher molecular weight residue gets iso-topically heavier. Thus the C_{15+} fraction of maturing crude oils gets progessively heavier as natural gas is formed (Figure 6.8).

Time

Age is much less important than thermal history in influencing petroleum maturation but, like any other kinetically controlled reaction, maturation will occur in less time (i.e. in younger reservoirs) if temperature is increased.

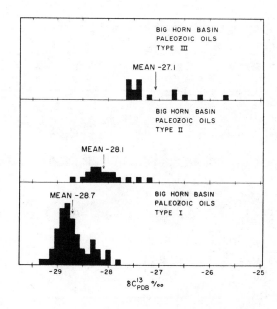

Figure 6.8: Changes in the distribution of δC^{13} values with increasing oil maturity in the Big Horn basin (Orr, 1974).

Natural deasphaltening

The steady increase in the concentration of the light ends which accompanies maturation may lead to precipitation of asphaltenes in the

95

reservoir and produce marked changes in the chemical and physical properties of the crude (as reflected by API gravity, etc.). As the heavy ends precipitate they take with them sulfur, nitrogen and metallic compounds and leave the producible oil lighter and of higher quality. The process of natural deasphaltening is exactly analogous to that employed in petroleum refineries to process heavy crudes (Table 6.2). Geological examples, such as the Keg River oils in the West Canadian Basin, have been described by Milner et al. (1977). They include cases of natural deasphaltening caused by influx of natural gas from outside the reservoir.

Table 6.2: Changes in chemical composition which accompany deasphaltening of selected oils (Billion et al., 1977).

	MURBAN	ARABIAN LIGHT	BUZURGAN	BOSCAN
Original oil				
Sp. gr.	0.982	1.003	1.051	1.037
% S	3.03	4.05	6.02	5.90
N ppm	3,180	2,875	4,500	7,880
Ni ppm	17	19	76	133
V ppm	26	61	233	1264
Asph. wt%	1.2	4.2	18.4	15.3
Oil deasphaltened with propane				
Sp. gr.	0.924	0.933	0.945	0.953
% S	1.80	2.55	3.60	4.70
N ppm	800	1,200	930	1,600
Ni ppm	1	1.0	2.0	6.0
V ppm	1	1.4	4.0	10.0
Asph. wt%	0.05	0.05	0.05	0.05

ALTERATION: CAUSES AND EFFECTS

Water washing

Several external factors may operate to produce changes in the composition of crude oils. Most of these are associated with moving

water because this acts as the transporting medium for materials being
removed from the crude oil or being brought into contact with it. Water
which flows past a reservoired crude oil will remove the most soluble
compounds preferentially. These include the lower molecular weight
hydrocarbons, particularly the aromatics, so that the water-washed
oil becomes much heavier. Water washing is only effective in changing
crude oil composition at the oil-water contact and the heavy, degraded
oil tends to be localized at the contact while the oil higher in the
trap may remain unaltered. Many tar layers (such as those which occur
in a large number of Middle East fields) have been produced in this way.

Bayliss (quoted by Milner et al. (1977)) studied water washing in
the laboratory and flushed a Gippsland basin oil with distilled water
for six months. Analysis of saturate materials in the residual oil and
the water showed (as expected) the preferential removal of the more
soluble naphthenes.

Bacterial degradation

Moving water is important not only for the things it removes from
crude oil but also for the materials it brings in. The more important
ones include bacteria and the nutrients and oxygen necessary for their
survival.

Most types of organic matter, including the compounds in crude
oils, can be affected by bacteria, but certain conditions must be satis-
fied if the bacteria are to survive. These include:

i. Water - bacteria require water as a medium to live in; they
    do not live in the crude oil itself. For this reason bacterial
    degradation of crude oils is concentrated at the oil-water
    contact.

ii. Nutrients - the inorganic nutrients (such as nitrate and phosphate) are brought in by the circulating water.

iii. Oxygen - the bacteria that are most effective in degrading oil are aerobic and hence require a supply of oxygen. This is brought in by the moving waters which are usually in hydraulic connection with the surface. If the moving water stops the available oxygen is quickly consumed and the bacteria die.

iv. Food - the crude oil or other organic matter is the food consumed by the bacteria.

v. Temperature - bacteria remain active at temperatures up to about $140^{O}F$ but can survive in a "dormant" state up to about $190^{O}F$.

vi. Toxins - one of the important roles of the water is to remove toxins, such as hydrogen sulfide, so that they do not reach concentration levels which are lethal to the bacteria.

Notice that all these conditions are met at the surface and here bacterial degradation of crude oils is very rapid and goes to completion. Oil seeps frequently provide examples of badly degraded crude oils.

Effects of Bacterial Degradation

Although almost all types of organic matter may be altered by bacteria the rates vary widely for different compound types and the effects of bacteria on crude oils produce some very specific changes in composition:

```
        n-paraffins            most susceptible to degradation
        aliphatic side chains
        branched paraffins                  |
        cyclic paraffins                    |
        aromatics                           ↓
        sulfur compounds       least susceptible
```

This highly selective removal of certain types of hydrocarbons produces

characteristic changes in crude oils as bacterial degradation progresses.

In general degraded oils have:

a) decreased n-paraffins. These are the preferred food for bacteria

   and are the first materials to be consumed.

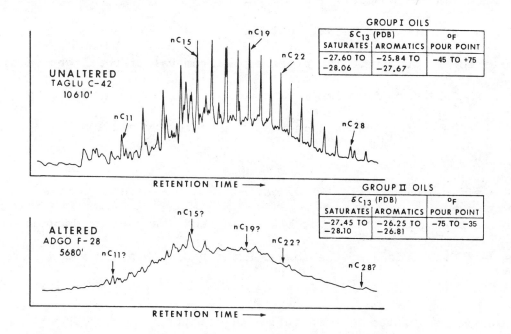

Figure 6.9: Whole oil chromatograms, the range of carbon
isotope ratios for saturate and aromatic fractions, and pour
points for typical unaltered and altered oils from the
Mackenzie Delta (Burns et al., 1975).

b) increased sulfur content. Percent sulfur increases as the paraffins

   are removed since sulfur compounds are resistant to bacterial

   degradation. Also, the bacteria produce sulfur compounds as part

of their biochemical processes and these metabolic products
are added to the crude oil.

c) increased nitrogen content. The reasons for the nitrogen increases
are the same as for sulfur - resistance of nitrogen compounds to
degradation and the addition of nitrogen compounds, such as proteins, from the life processes of the bacteria.

d) increased optical activity. Removal of compounds with no optical
activity (such as the normal paraffins) leaves the oil with enhanced optical activity. Also many of the compounds produced by
the bacteria are optically active and find their way into the crude
oil giving increased optical activity.

e) decreased API gravity. The preferential removal of the light ends
and high $^{\circ}$API paraffins leaves a crude oil enriched in the heavy
ends with a corresponding higher specific gravity and lower API
gravity.

The effects of bacterial degradation have now been recognized in
oil fields throughout the world, but the effects were first documented
in the Bell Creek field of the Powder River basin (Winters and Williams,
1969). Biodegradation is most severe in the central part of the elongate
field and is associated with an influx of fresher water. API gravities have
been reduced to 32.5°, which is 10° lower than for the unaltered crudes
at the end of the barrier sand reservoir. Other significant changes in
the oils are summarized in Table 6.3 and Figure 6.10.

Deroo et al. (1975) published data for a group of Canadian oils
showing various degrees of bacterial alteration (Figure 6.11). Normal
paraffins over the whole range studied (C_{15}-C_{30}) were removed first and

Table 6.3: Data for selected Bell Creek Oils with various
amounts of bacterial degration.

Sample	Extent of n-paraffin loss	% Nitrogen	Specific Rotation (Aliphatic fraction)
M8-97	All	0.14	1.05
M8-98	All	0.15	0.86
M8-104	All	0.14	0.88
M8-107	below C_{18} only	0.14	0.88
M8-108	below C_{18} only	0.14	0.78
M8-102	partial loss, entire range	0.09	0.46
M8-99	none	0.05	0.48

the isoprenoids pristane and phytane became relatively more abundant,
but even they were lost from the most severely degraded oils. Multi-
ringed naphthenes were resistant to degradation and their relative
abundances remained unchanged. This sequence of changes in composition
as biodegradation proceeds has been duplicated in laboratory studies
(Bailey et al., 1973a).

The study of crude oils from the northeast part of the Williston
basin by Bailey et al. (1973) is a particularly well documented example
of the effects of bacteria in degrading crude oil. Again, the most
severe degradation is associated with the fresher waters. The low salinity
suggests a meteoric source and this was confirmed by isotopic data.
Figure 6.12 shows that the High Prairie oil has been badly degraded so
that few normal paraffins remain although isoprenoid compounds are
prominant. This oil is associated with waters containing less than
50,000 mg/liter of dissolved solids, has a low API gravity and a sulfur
content >3%, among the highest in the area studied. In contrast the
Staughton oil is associated with saline waters, has a full suite of

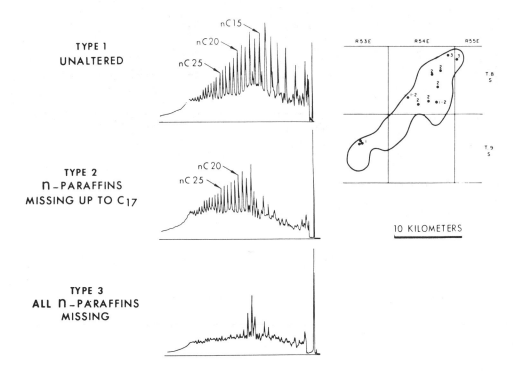

Figure 6.10: Location of sample collection points in the Bell Creek
field and gas chromatograms of the unaltered, partly altered and
extensively altered oils (from Winters and Williams, 1969).

normal paraffins, a higher API gravity and a lower sulfur content.

SUMMARY

Changes in the chemical composition of a crude oil which occur in

the reservoir due to maturation and alteration have been well reviewed

by Milner et al. (1977). These changes are summarized in Figure 6.13

(Evans et al., 1971).

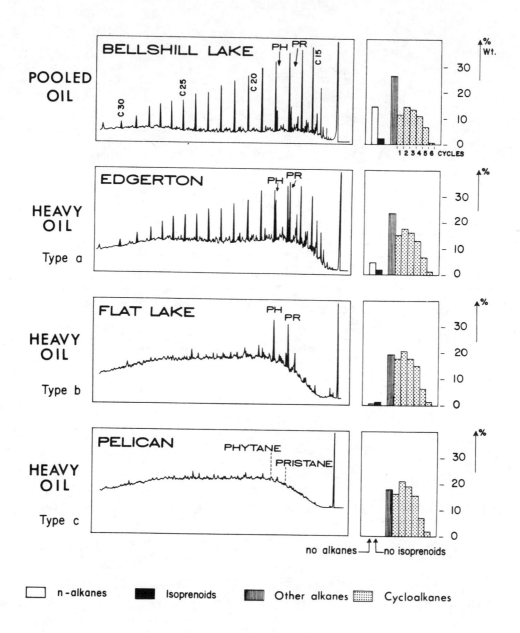

Figure 6.11:
Gas chromatograms and compositional data for
Canadian oils showing the progressive changes produced by
bacterial degradation. (Unaltered at the top; most
severely biodegraded at the bottom) (Deroo et al., 1975).

SATURATED HYDROCARBONS

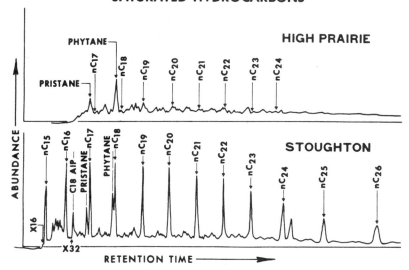

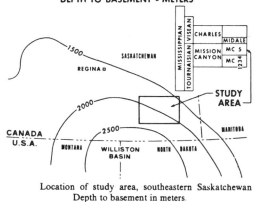

Location of study area, southeastern Saskatchewan
Depth to basement in meters.

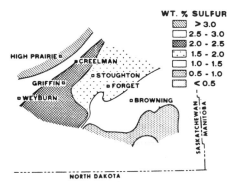

Geographic variations in sulfur contents of Mission
Canyon oils of Saskatchewan. Locations of town of Weyburn
and of sources of six samples are shown.

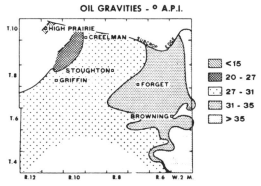

Salinity of MC5 waters.

API gravity variations within MC5 oils of Saskatchewan. Locations of sources of six oil samples selected for detailed analysis are shown.

Figure 6.12: Bacterial degradation and its effects as
illustrated by crude oils from S. E. Saskatchewan, Canada
(Bailey et al., 1973).

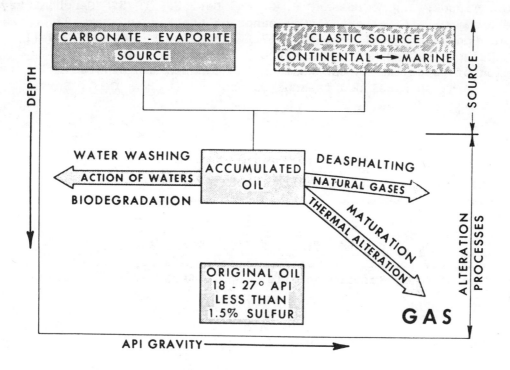

Figure 6.13

REFERENCES

Bailey, N. J. L., Jobson, A. M. and Rogers, M. A., 1973.   Bacterial degradation of crude oil: comparison of field and experimental data: Chem. Geol., v. 11, p. 203-221.

Bailey, N. J. L., Krouse, H. H., Evans, C. R. and Rogers, M. A., 1973. Alteration of crude oil by waters and bacteria - evidence from geochemical and isotope studies: Bull. AAPG, v. 57, p. 1276-1290.

Barton, D. C., 1934. Natural history of the Gulf Coast crude oil: in W. E. Wrather and F. H. Lahee (Eds.) "Problems of Petroleum Geology", Sidney Powers Memorial Volume, AAPG, p. 109-155.

Billon, A., Peries, J. P., Fehr, E. and Lorenz, E., 1977. SDA key to upgrading heavy crude: O. & G.J., Jan. 24th p. 43-48.

Burns, B. S., Hogarth, J. T. C. and Milner, C. W. D., 1975. Properties of Beaufort Basin liquid hydrocarbons: Bull. Canad. Petro. Geol., v. 23, p. 295-303.

Connan, J., LeTran, K. and Van der Weide, B., 1975. Alteration of petroleum in reservoirs: Proc. 9th World Pet. Congr. v. 2, p. 171-178.

Deroo, G., Tissot, B., McCrossan, R. G. and Der, F., 1975. Geochemistry of the heavy oils of Alberta: Oil sands of the Future, Mem. #3, Canadian Soc. Petr. Geol., p. 148-167. (See also Deroo, G., Powell, T. G., Tissot, B. and McCrossan, R. G., 1977. The origin and migration of petroleum in the western Candian sedimentary basin, Alberta. A geochemical and thermal maturation study: Geol. Surv. Canada, Bull. 262, 136 pp.)

Dobryanski, A. F., 1961. Chemistry of Petroleum: Gostoptekhizdat, Leningrad, 450 pp.

Evans, C. R., Rogers, M. A. and Bailey, N. J. L., 1971. Evolution and alteration of petroleum in western Canada: Chem. Geol., v. 8, p. 147-170.

Koons, C. B., Bond, J. G. and Peirce, F. L., 1974. Effects of depositional environment and postdepositional history on chemical composition of Lower Tuscaloosa oils: Bull. AAPG, v. 58, p. 1272-1280.

Landes, K. K., 1967. Eometamorphism, and oil and gas in time and space: Bull. AAPG, v. 51, p. 828-841.

McNab, J. G., Smith, P. V. and Betts, R. L., 1952. The evolution of petroleum: Ind. Eng. Chem., v. 44, p. 2556-2563.

Milner, C. W. D., Rogers, M. A. and Evans, C. R., 1977. Petroleum transformation in reservoirs: J. Geochem. Expl., v. 7, p. 101-153.

Orr, W. L., 1974. Changes in sulfur content and isotopic ratios of sulfur during petroleum maturation - study of Big Horn basin Paleozoic oils: Bull. AAPG, v. 58, p. 2295-2318.

Parker, C. A., 1973. Geopressures in the deep Smackover of Mississippi: J. Petrol. Tech., August, p. 971-979.

Pusey, W. C., 1973. How to evaluate potential oil and gas source rocks: World Oil, April, p. 71-74.

Rogers, M. A., McAlary, J. D. and Bailey, N. J. L., 1974. Significance of reservoir bitumens to thermal maturation studies, Western Canada Basin: Bull. AAPG, v. 58, p. 1806-1824.

White, D., 1915. Some relations in origin between coal and petroleum: J. Wash. Acad. Sci., v. 5, p. 189-212.

Williams, J. A. and Winters, J. C., 1969. Microbial alteration of crude oil in the reservoir: in Petroleum transformation in geologic environments - symposium, Amer. Chem. Soc. 158th Nat'l. Mtg., New York. Paper No. PETR 86:E22-E31.

TOPIC 7

SOURCE ROCKS

INTRODUCTION

We turn our attention next to the source rocks where petroleum
generation has occurred. It is important to be able to recognize these
rocks, particularly in the early stages of exploration in a new venture
area because if more than one source rock is present the area is much
more attractive. An estimate of how prolific the source has been and
some indication of the nature of the products (oil or gas) is also
valuable.

The source rocks available for examination contain the hydrocarbon
residue left after some of the organic material has been removed (by a
migration mechanism which is not fully understood). Since only a few
percent of the hydrocarbons generated in the source rock migrate presumably
there is only a small change in the chemical composition of the source
rock extract. If migration occurs in aqueous solution, for example, the
source rock should be enriched in the least soluble components. However,
an enrichment can only be determined relative to the orginal composition
and this is not normally known. In practice the methods currently in
use for identifying source rocks are methods for detecting generation.
It is then assumed that migration has occurred.

Use of the term "source rock" is often ambiguous and may be applied
to rocks which are in very different stages of generation and expulsion.
Dow (1977) has provided a list of definitions "to alleviate the confusion
which exists in the geochemical literature" and a slightly modified

Table 7.1: Definitions pertinent to source rocks.
(Modified after Dow, 1977).

SOURCE ROCK - A unit of rock that has generated and expelled oil
or gas in sufficient quantity to form commercial accumulations.
The term "commercial accumulations" is by definition, variable.

LATENT SOURCE ROCK - A source bed that exists but is as yet con-
cealed or undiscovered. Usually refers to unexplored areas or
deep portions of developed basins.

POTENTIAL SOURCE ROCK - A unit of rock that has the capacity to
generate oil or gas in sufficient quantities to form commercial accumulations but has not yet done so because of insufficient thermal maturation.

ACTIVE SOURCE ROCK - A source bed that is in the process of
generating oil or gas.

SPENT SOURCE ROCK - A source bed that has completed the process
of oil or gas generation and expulsion. A source bed may be
spent for oil and active for gas.

INACTIVE SOURCE ROCK - A source bed that was once active but has
temporarily stopped generating prior to becoming spent.
Usually associated with thermal cooling due to uplift and
erosion. Still has some potential left.

LIMITED SOURCE ROCK - A unit of rock that contains all the pre-
requisites of a source bed except volume. Commonly refers
to thin shale laminations in carbonates or thin coals in
continental deposits.

---

version of this is given here as Table. 7.1.

For many years the oil industry rule of thumb has been that "black

shales" are source rocks. There is at least a grain of truth in this

because red beds, for example, usually indicate oxidizing conditions

and imply low organic matter contents. There are many source rocks which

are not black and some black shales which contain little organic matter

- the color being due to pyrolusite (MnO_2) or other inorganic material.

Trask and Patnode (1942) made the first extensive study of source rocks and found no correlation of source character with color. They also failed to find any relationship with organic matter content and chemical composition although carbon/nitrogen ratio did show some promise. Their approach of selecting rocks presumed to be sources from their geological associations and then doing chemical analyses in an attempt to establish some source rock characteristic has been used in many subsequent studies.

SOURCE ROCK INDICATORS

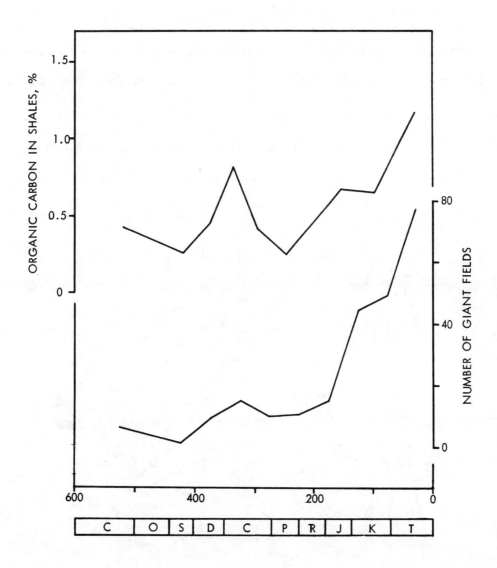

Figure 7.1:
Variation of organic carbon content of shales with geologic age
compared with numbers of giant fields (modified from Barker, 1977).

It now seems that an evaluation of a shale, or other rock type, for source rock character must provide answers to the following questions- (i) does the rock have sufficient organic matter? (ii) is the organic matter of the right type? (iii) has this organic matter generated petroleum? and (iv) has the generated petroleum migrated out? These questions will be discussed individually.

Does the rock have sufficient organic matter?

The relationship between high organic matter content in the rocks and occurrence of reservoired petroleum is well documented both globally (Figure 7.1) and regionally (Figure 7.2). Statistics for giant fields worldwide show a remarkably good correlation with the available organic material as indicated by the average organic carbon content of sedimentary rocks. In local areas this trend also applies and reservoired

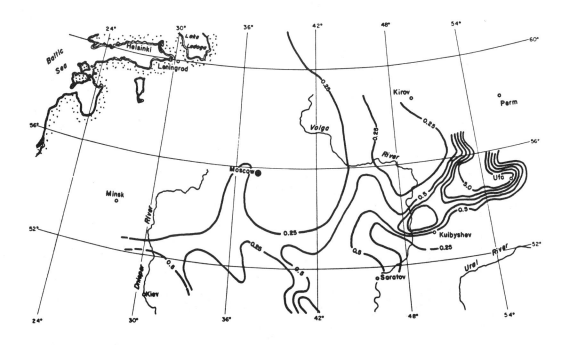

Figure 7.2:
Distribution of organic carbon in Upper Devonian shales of the
Russian platform (from Ronov, 1958).

110

petroleum is frequently associated with regions of higher than average organic carbon contents. Although some source rocks have carbon contents as high as 10% most are between 0.8% and 2.0% and it seems to require a minimum of about 0.5% active organic matter (0.4% organic carbon) in shales (but rather less in carbonates) to produce a source rock (Dow, 1977). To some extent low organic matter content can be offset by larger source rock volumes or the presence of a type of organic matter that is a prolific generator.

Is the organic matter of the right type?

It is obvious that the organic matter in the potential source rock must be of a type which is capable of generating petroleum. Recycled material, which has already been buried and is now no longer capable of generating, should be excluded from the total when establishing the percentage of organic matter in the rock. Although recycled material can often be recognized visually or by its high vitrinite reflectance values (Figure 7.3) it is usually difficult to estimate the weight

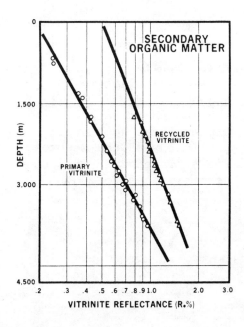

Figure 7.3: Differentiation between primary and recycled vitrinite using vitrinite reflectance (Dow, 1977).

percent present. Organic matter type has an important influence on the nature of the petroleum products. This is discussed below.

Has the organic matter generated petroleum?

Overall control of generation depends on temperature modified by time. While depth and age often provide a rough guide they must be used with caution since depth of generation will depend on the local geothermal gradient. In the Niger delta two wells 200 miles apart have geothermal gradients of $1.0^{\circ}F/100$ ft. and $3.0^{\circ}/100$ ft respectively (Nwachukwu, 1976). Reference to Figure 6.6 (page 93) shows that the hotter well has passed through the oil window at 7500 ft while in the other well generation has not even started at this depth. Geothermal gradients in the past may have been very different from present day values. In some cases the paleogeothermal gradients can be inferred from the regional setting (e.g. high in rift zones, low in deltas) or from vitrinite reflectance and electron spin resonance ("esr") studies. The vitrinite reflectance data of Figure 7.4 shows a break at the Tertiary/Mesozoic unconformity with different

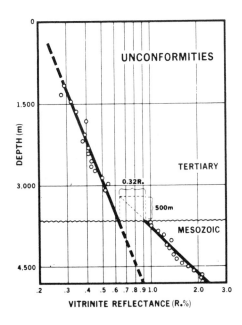

Figure 7.4: Vitrinite reflectance data for an Indonesian well (Dow, 1977) showing that there was a higher local geothermal gradient when the Mesozoic rocks formed than when the Tertiary ones formed.

112

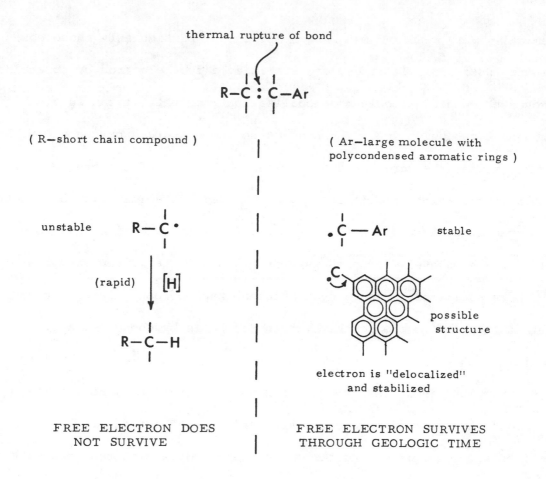

Figure 7.5: Schematic representation of the events
following thermal bond rupture.

slopes above and below, indicating that the local geothermal gradient

was higher when the Mesozoic rocks formed than when the Tertiary ones

formed.

The esr techniques has not been widely used but does show consid-

erable promise because of its potential application to all types of

organic matter (Ho, 1977), while reflectance methods are restricted to

vitrinite which may be absent from key rock units. In the esr technique

the concentration of free electrons in the organic matter is determined.

These are produced as "free radicals" when a chemical bond is thermally

broken as temperature rises (Figure 7.5). Small free radicals are chemically very reactive but larger molecules, particularly those containing fused aromatic rings, can stabilize a free electron by spreading it over the whole molecule ("delocalization") so that it can survive throughout geologic time. The higher the temperature experienced by the organic matter the more bonds that will be broken, and the greater the number of free electrons in the organic matter. Stabilization of electrons depends critically on organic matter type, and the technique has to be calibrated separately for each type of organic matter. Few examples have been published since Pusey's original paper (Pusey, 1973) but esr has been applied to the northern North Sea basin by Cooper et al. (1975) who infer that maximum paleotemperatures were reached in the Tertiary. Other examples, including the COST B-2 well, are given by Ho (1977).

Both physical and chemical characteristics of kerogen reflect the combined effects of time and temperature and indicate thermal maturity. With increasing time-temperature kerogen color intensifies and darkens (Figure 4.5, page 46); vitrinite reflectance increases (Figure 4.6, page 47); palynomorph translucency decreases; thermal stability (as indicated by pyrolysis) increases; and the elemental composition of kerogen shows a progressive increase in percent carbon.

The values corresponding to the onset of generation are given in Table 7.2, and Figure 7.6 shows how some of them are related to peak generation of oil and gas. As an example of the geological application of kerogen color in defining geochemical facies Figure 7.7 shows contours of equal color intensity for kerogen in the Paris and Aquitaine basins. Here crude oil occurs where the color rank is between 2 and 3, and gas occurs where it is over $3\frac{1}{2}$.

Table 7.2: Indicators of the onset of generation

Parameter	Values	Comment
C_1-C_5 (amount)	10^{-2} to 10^{-3} g/g of C	Claypool et al.(1977)
C_{15+} (amount)	500 ppm	Philippi (1965)
C_{15+}/T.O.C	0.01	
CPI	approaches 1.0	Brayand Evans (1961)
kerogen color	darker than yellow-orange (2-)	Staplin (1969)
vitrinite reflectance	$0.45R_O$	Dow (1977): $0.60R_O$ for peak generation
H/C in kerogen H/C vs 0/C in kerogen		LaPlante (1974) varies with organic matter type: Tissot et al. (1974)
%C in kerogen	75%	LaPlante (1974)
palynormorph translucency		Grayson (1975): calibrated only for carya
pyrolysis yield		Claypool and Reed (1976)
pyrolysis temperature	460^OC	depends on heating rate. Claypool
electron spin resonance (esr)	complex	Pusey (1973), Ho (1977)

Although kerogen accounts for the major part of the organic matter in rocks it is the bitumens which eventually accumulate to form reservoired petroleum, and the most direct way of establishing the amount and composition of generated material is to analyze the solvent-extractable

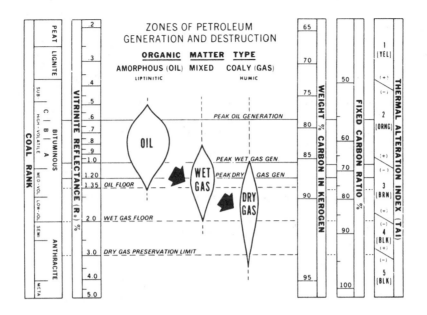

Figure 7.6: Correlation of the coal rank scale with various maturation indices and the zones of petroleum generation and destruction. The relative importance of petroleum generation zone depends on the composition of the original kerogen (Dow, 1977).

bitumens. The onset of generation produces an increase in the ratio (amount of bitumens/total organic matter). As the amount of bitumens increases their chemical composition changes because the non-biogenic thermal processes act to remove biological characteristics such as odd chain length predominance in the normal paraffins, the high abundance of the 4- and 5-ringed naphthenes and the relative amount of the isoprenoids (Figure 7.8). In addition to the C_{15+} extract the quantity and composition of the light hydrocarbons also gives useful information and they have been used successfully for defining the mature facies in the western Canadian basin (Evans and Staplin, 1971) and in the Sverdrup basin (Snowdon and Roy, 1975) (Figure 7.9).

Recently pyrolysis techniques have been used to evaluate source rocks (Barker, 1974; Claypool and Reed, 1976; Espitalie et al., 1977).

116

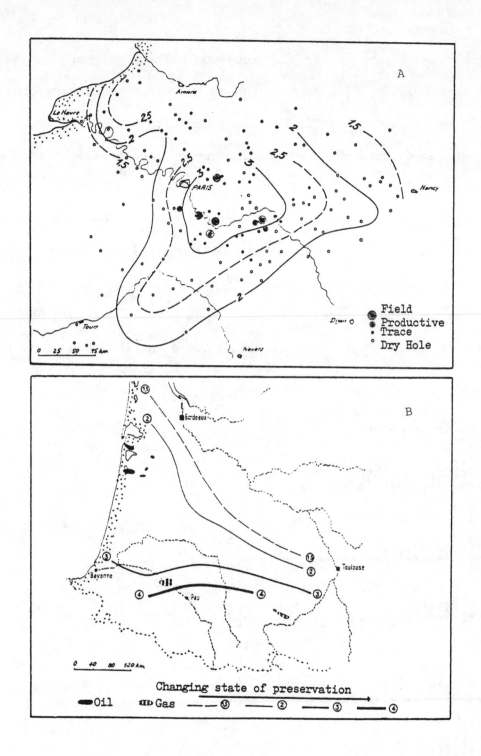

Figure 7.7: Lias of the Paris basin (A) and Jurassic of the
Aquitaine basin (B) contoured with organic matter color (1 -
pale yellow; 5 - black) (after Correia, 1971).

In one common procedure rock samples are heated at a steadily increasing

temperature (e.g. 20° C/min.) and the total evolved hydrocarbons monitored as a function of temperature. Hydrocarbon release curves generally show a peak around 200°C which is produced by hydrocarbons already present in the rock due to generation in the subsurface (Peak 1), and a second peak at higher temperature which can be assigned to the breakdown products of the kerogen (Peak 2). Thus, Peak 1 is comparable to the solvent extractable hydrocarbons and Peak 2 to the total organic carbon. As rocks are buried more deeply in the subsurface and subjected to higher temperatures increasing amounts of hydrocarbons are generated and Peak 1 gets larger. This increase occurs at the expense of Peak 2 which not only decreases in size, but also moves to higher temperatures as the less thermally

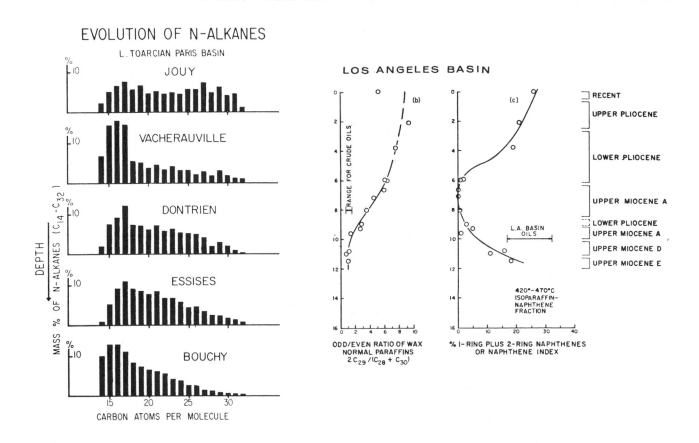

Figure 7.8: Changes in the chemistry of rock extracts with depth
(Los Angeles basin - Philippi, 1965;
Paris basin - Tissot, et al., 1971).

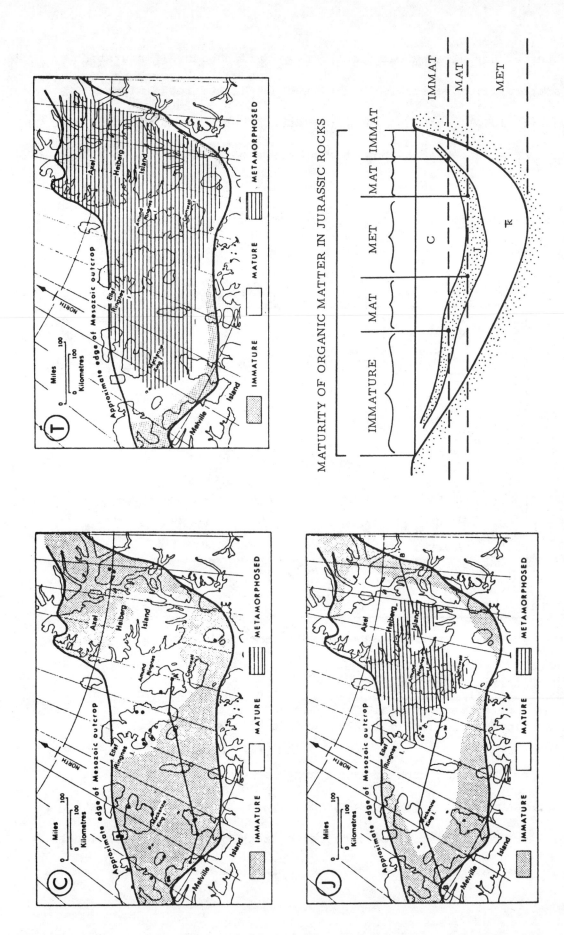

Figure 7.9: Facies of organic metamorphism in Mesozoic rocks of the Sverdrup basin (After Snowdon and Roy, 1975).

stable material has already been broken down during natural maturation leaving a thermally more stable kerogen residue in the rock (Figure 7.10) (Barker, 1974). Both the ratio of (Peak 1/Peak 2) and the temperature of the maximum in the hydrocarbon release curve for Peak 2 can be used to indicate maturity (Figures 7.11, 7.12). The hydrocarbons in Peaks 1 and 2 can be analyzed by suitable trapping procedures coupled with gas

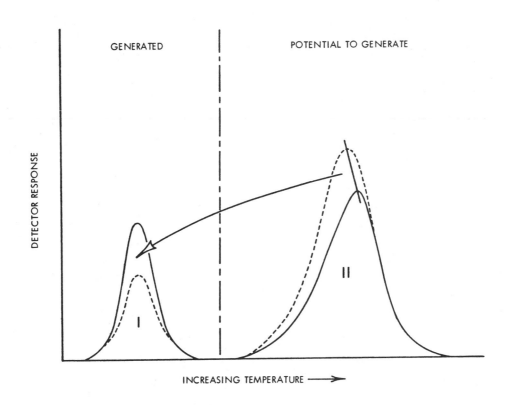

Figure 7.10: Schematic representation of way in which low-temperature peak (Peak 1) increases with depth at expense of high-temperature peak (Peak 2). Dashed line response for shallow sample; solid line, response for deep sample.

chromatography. The character of these hydrocarbons, particularly the relative abundances of short and long chains, can be a useful indication of the oil- or gas-generating ability of the rock. Pyrolysis techniques show great promise, because separation of organic matter from the rock

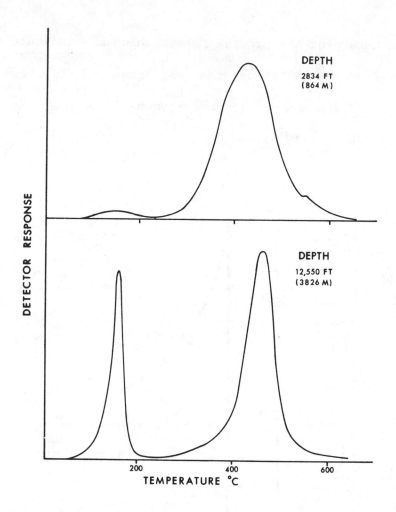

Figure 7.11: Detector response - temperature curves for two shale
samples from different depths in the same well
(Barker, 1974).

is not necessary, analysis takes only about an hour, and because only

small samples are required (typically around 50 mg.) so that cuttings

provide adequate material.

Has the generated petroleum migrated?

Of all the questions pertinent to source rock evaluation this is

the most difficult to answer because generally-applicable experimental

procedures have not been developed. Correlation of the source rock

extract to a reservoired crude oil provides the only direct indication

that migration has occurred. Other information which can be helpful

121

includes (a) the nature of the mineral matrix which may indicate availability of water from mineral transformation (Johns and Shimoyama, 1972) and (b) comparison of the composition of bitumens in open and isolated "microreservoirs" (Barker, 1974).

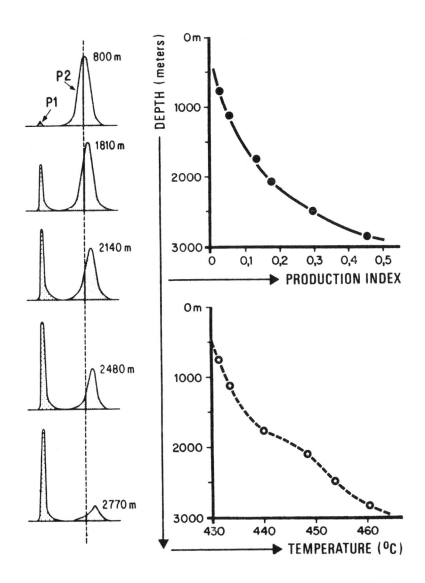

Figure 7.12:
Rapid characterization of the degree of evolution of source
rocks in sedimentary section of tertiary age
(Espitalie et al., 1977).

CASE STUDY: COST B-2 WELL

The application of organic geochemical analyses to the evaluation of a frontier area is well illustrated by the COST* B-2 well drilled off the northeast coast of the U.S.A. in the Baltimore Canyon prior to a recent lease sale. The well was drilled through a sand-shale sequence (with minor limestone, coal and lignite) to a total depth of 16,043 ft. Biostratigraphic studies show that 5000 ft. of Tertiary and Quaternary age sediments overlie 3000 ft. of Upper Cretaceous and 8000 ft. of probably Lower Cretaceous sediments (Scholle, 1977). The geochemical studies were carried out by the U.S.G.S., service companies and participating oil companies using a wide range of techniques (Table 7.3, Figure 7.13). The results showed that rocks containing more than 1% organic carbon were present from 3-6,000 ft. and from 10-14,000 ft, but that the whole section penetrated was immature. This conclusion was based on the low values for C_{15+}/org. carbon and the high CPI values combined with low vitrinite reflectance, low temperatures of maximum pyrolysis yield and the presence of kerogens with high H/C ratios and low percent carbon. Isotopic data, kerogen color, and the composition of the C_4-C_7 fraction were consistent with immaturity. However, Claypool et al. (1977) concluded that although there were no fully mature source rocks in the section the deeper samples were almost at the point where crude oil generation becomes significant. Generation in laterally equivalent deeper units is possible. Also C_1-C_7 analyses suggest incipient wet gas generation in the deeper part of the well.

---

* Continental Offshore Stratigraphic Test

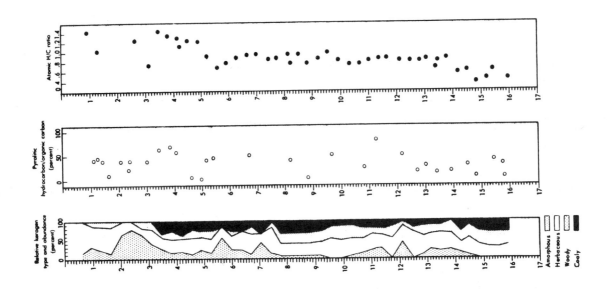

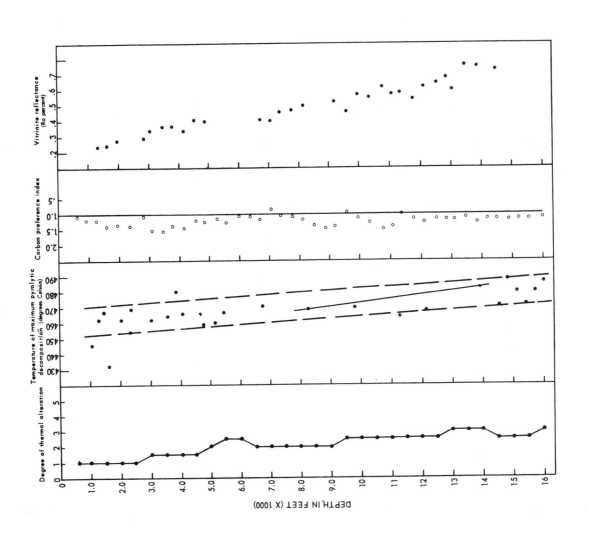

Figure 7.13: Geochemical data from the COST B-2 well (Scholle, 1977).

This study is a good example of combining several geochemical techniques, but in many cases this must be done with care because many companies have developed self-consistent criteria which are strictly valid only for data generated by their own analytical procedures.

Table 7.3: Geochemical techniques used to study cutting from the COST No. B-2 well, U.S. Mid-Atlantic Outer Continental Shelf Area

TECHNIQUE	ANALYZING GROUP
$C_1 - C_7$	U.S.G.S.
$C_4 - C_7$	Geochem
C_{15+}	U.S.G.S. Geochem
% organic carbon	U.S.G.S. Geochem
kerogen type	Geochem
kerogen color	U.S.G.S.
%C in kerogen	Amoco
H/C in kerogen	Amoco
vitrinite reflectance	Superior
carbon isotope values	Phillips
pyrolysis (thermal stability)	U.S.G.S.
pyrolysis (composition of products)	U.S.G.S.

OIL OR GAS? - IMPORTANCE OF ORGANIC MATTER TYPE

In many areas, particularly those remote from consumers, the economics of transportation make oil much more desirable than gas. The ability to use source rock character to predict the nature of the hydrocarbons generated is a valuable asset in exploration. Any organic matter will generate gas if buried deeply enough (see TOPIC 6), but at intermediate depths the type of organic matter exercises primary control over the nature of the petroleum products.

The composition of the organic matter incorporated in sediments reflects the character of the organisms which produced it and we can make a broad distinction between organisms wich grow on the land surface ("terrestrial") and those which grow in water ("aquatic"). Terrestrial organisms must support their own weight and protect themselves from dehydration due to evaporative loss of body fluids, but neither of these functions is necessary in an aqueous growth medium where the water prevents dehydration and gives support. Both the wood which provides structural support in terrestrial plans and the surface waxes which prevent evaporation are absent from aquatic organisms.

Wood is made up of the two biopolymers, cellulose and lignin, together with smaller amounts of lipids. Cellulose can be biodegraded but the lignin is much more resistant and survives deep into the sub-surface where rising temperature causes cleavage of the short side chains producing methane and ethane (Figure 7.13). The aromatic ring structures remain part of the kerogen. Thus, the kerogens derived from woody precursors are gas generating. Coals provide an extreme example of predominantly wood-derived kerogen where the organic matter accounts for the bulk of the rock. They can be source rocks and their role in producing much of the gas in the southern North Sea and in Holland (Sole, Leman, Gronigen, etc.) is well documented (Lutz et al., 1975). Gases produced from coal freqently have above average amounts of associated nitrogen.

Surface waxes, which prevent evaporation, are esters of long chain acids and alcohols and give very long chain n-paraffins when they are thermally degraded in the subsurface. They appear to be the only source of n-paraffins with chain lengths greater than about nC_{25} and crude oils containing high percentages of the long chain paraffins are waxy and have high pour points. These oils are generally associated with near-shore sedimentation especially in deltaic settings (Hedberg, 1968). The Altamont-Bluebell oils of the Uinta basin, with pour points over $100^{o}F$, provide a good example of waxy crudes derived from terrestrial organic matter. Sutton (1977) has correlated waxy crude oils in the Java Sea to their source rocks and reports that 90% of the organic matter is land derived and 79% of the total organic matter is made of the waxy coatings from plants ("cuticle").

Amorphous kerogen of presumed algal origin seems to generate the range of normal crude oils characterized by relatively low amounts of n-paraffins in the nC_{25+} range and a maximum around C_{19} and with low gas-oil ratios.

Organic matter type is most readily distinguished by microscopic examination and depth profiles showing variations in composition are widely used (see for example the COST B-2 well data in Figure 7.13). There are chemical differences between organic matter types, and ter-restrial material has a lower H/C ratio (at any given level of thermal maturity). This appears to be anomalous since it generates the most hydrogen-rich product (methane with H/C = 4), but the situation is readily explained by reference to the lignin structure. The hydrogen is low overall because of the high content of aromatic rings while the high hydrogen gas comes from cleaved short side chains. The nature of the

Figure 7.14

SOURCE OF ORGANIC MATTER	TYPE OF ORGANIC MATTER	ROLE	CHEMICAL STRUCTURE AND DEGRADATION PRODUCTS	PETROLEUM TYPE
TERRESTRIAL	LIGNIN (monomer)	STRUCTURAL SUPPORT	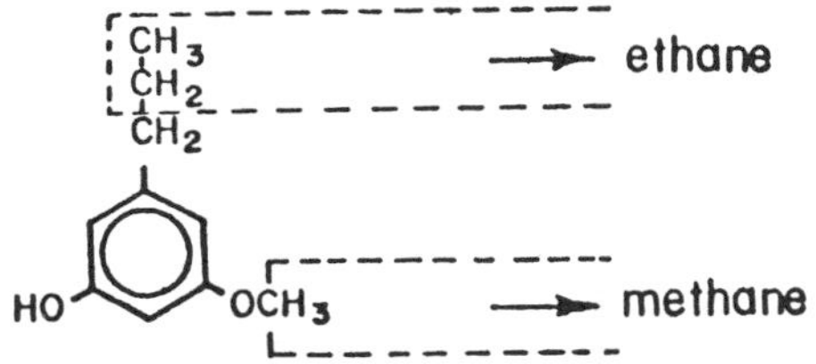 CH_3, CH_2, CH_2 → ethane; OCH_3 → methane	NATURAL GAS
TERRESTRIAL	SURFACE WAXES (ESTERS)	PREVENT EVAPORATION	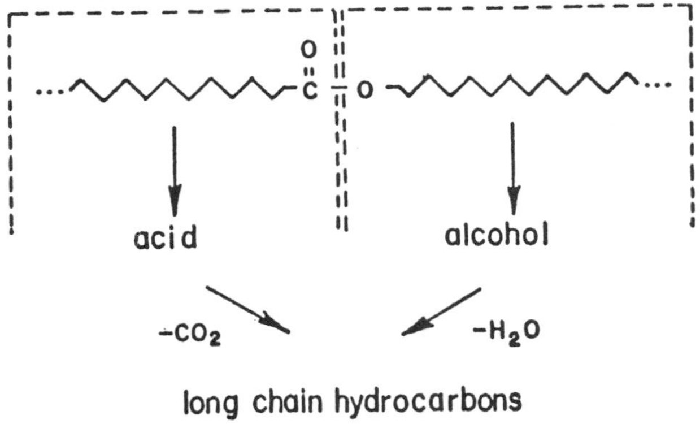 acid; alcohol; $-CO_2$; $-H_2O$; long chain hydrocarbons	WAXY CRUDE OILS
AQUATIC	LIPIDS		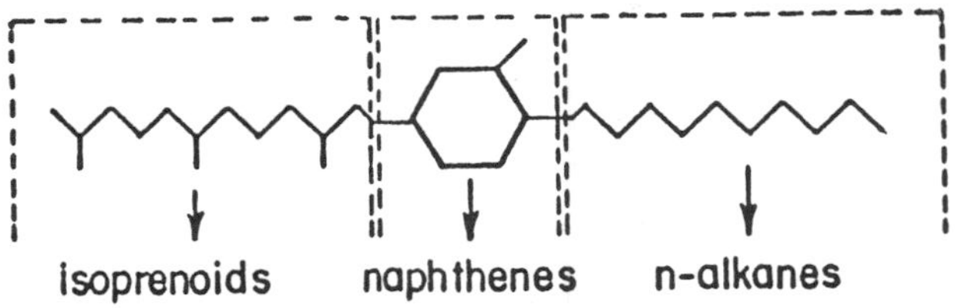 n-alkanes; naphthenes; isoprenoids	NORMAL CRUDE OILS

128

products of laboratory thermal degradation (pyrolysis) experiments
is also useful in showing the nature of the petroleum generated by the
source rock.

GEOLOGIC CONTROLS ON ORGANIC MATTER DISTRIBUTION

The distribution of various types of organic matter has economic
significance because terrestrially derived material tends to be gas
generating while aquatic organic matter is oil generating. The geographic

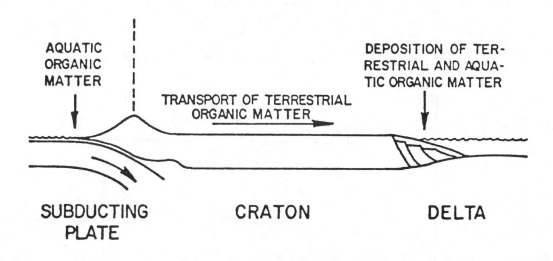

Figure 7.15

location of areas of high productivity for these two extreme types of
organic materials is very different. Aquatic organisms (mainly algae)
are abundant where nutrients (mainly nitrate and phosphate) are plenti-
ful and sunlight is available. The up-welling of deep cold waters
brings nutrients into the euphotic zone at high latitudes and stimulates
high productivities in these areas so that the seas off Antartica are
the most productive in the world. In contrast the most prolific production of terrestrial (especially woody) organic material is in the
equatorial rain forests, all of which are located within 15° of the

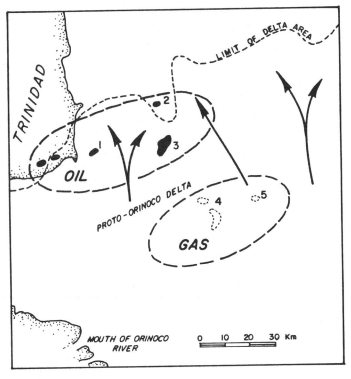

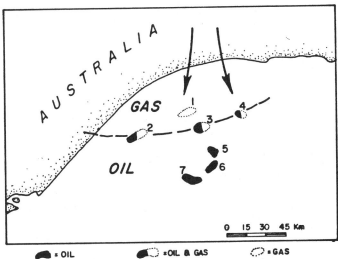

Figure 7.16: Distribution of oil and gas relative to
paleoshorelines in deltaic sequences. The arrows indicate
direction of sediment input.
Top - Proto-Orinoco delta, Middle Miocene Cruse Formation.
1. Galeota, 2. Samaan, 3. Teak, 4. Southwest Galeota,
5. Queens Beach.
Bottom - Gippsland Basin, Eocene LaTrobe Formation.
1. Snapper, 2. Barracouta, 3. Marlin, 4. Tuna, 5. Halibut,
6. Mackerel, 7. Kingfish.

equator. There are no forests in Antartica!

Virtually all sediments containing organic matter are initially deposited in water, but terrestrial organic matter grows on the land surface and must be transported to the area of deposition. Rivers and the associated drainage areas exert a major influence on the transfer of terrestrial organic matter and control the types of environments in which terrestrial organic matter is deposited (Figure 7.15). Not surprisingly tertiary deltas are a major depository for land-derived organic materials and they also contain disproportionately high amounts of gas compared with other depositional environments (Halbouty et al., 1970).

For several reasons the distribution of terrestrial and aquatic organic material in deltas is not uniform. Terrestrial material has its

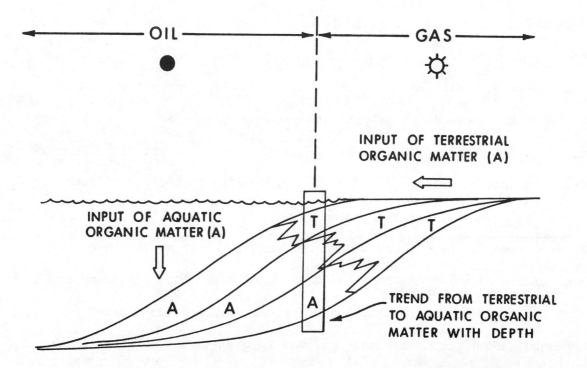

Figure 7.17: Schematic representation of the distribution of terrestrial and aquatic organic matter in a prograding delta.

highest concentration nearshore, at the point of input, and is steadily diluted by aquatic material with increasing distance from shore. Growth of aquatic organisms is stimulated by the nutrients brought down by the river. Although nothing has been published on the hydraulic behaviour of various types of organic matter we can anticipate that the chunky and compact woody material will be deposited with slightly coarser sediments than the "fluffy" aquatic material. Again, this is a trend for terrestrial nearshore and aquatic offshore. This distribution of organic matter type should be reflected in the distribution of oil and gas since one type is gas prone and the other oil prone. Such a trend has been observed. In the Miocene Cruse Formation of the proto-Orinoco delta sediments (Michelson, 1976) all the gas fields are nearer to the shore line while the oil fields are located farther to the north which is farther away from the old shore line (Figure 7.16). The same situation is observed in the Eocene deltaic sediments of the Latrobe Formation in the Gippsland Basin between Australia and Tasmania. The delta prograded southward and the most nearshore (northerly) fields produce gas (e.g. Snapper), there are then a series of oil-and-gas fields (e.g. Marlin) and farthest from shore lie the oil fields (e.g. Mackerel, Kingfish, etc.). The depth range for these oil and gas fields is only about 2000 ft. - insufficient for the trend to be a thermal one produced by differing depths of burial. Many of the oils in the Gippsland Basin have high pour points and are waxy, which is another indication of terrestrially sourced material (Hedberg, 1968). This is also observed in the Mahakam delta (Magnier et al., 1975) where the oils farther from shore (e.g. Bekapai) are normal while nearer to the paleoshoreline is gas and waxy crude oil (Handil, for example, has a pour point of $85^{o}C$).

With time deltas prograde seawards, and terrestrial organic material
is deposited over the previously deposited aquatic material (Figure
7.17) producing a trend from gas generating to oil generating with
depth. This sequence was documented by Dow and Pearson (1975) for the
Mississippi delta (Figure 7.18). The effects of such a distribution can
be seen in the Marlin field of the Gippsland Basin which has six productive sands with the top five containing gas and the deepest one oil.

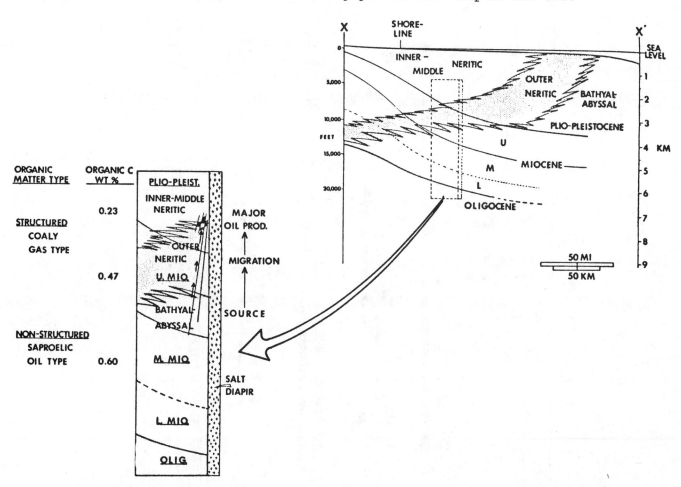

Figure 7.18: Profile of organic matter type in the Mississippi delta
(Dow and Pearson, 1975).

The vertical sequence of organic matter types is just the opposite
on divergent continental margins ("pull-aparts") because the oldest and
deepest sediments were produced in a continental rift and have the

highest percentage of terrestrially derived organic matter. They are

overlain by sediments containing increasing contributions from marine

organic matter. This sequence of sediment types from continental,

through taphrogenic and marginal sea, to open marine is well documented

offshore Brazil (Campos et al., 1975). Here the deep Lower Cretaceous

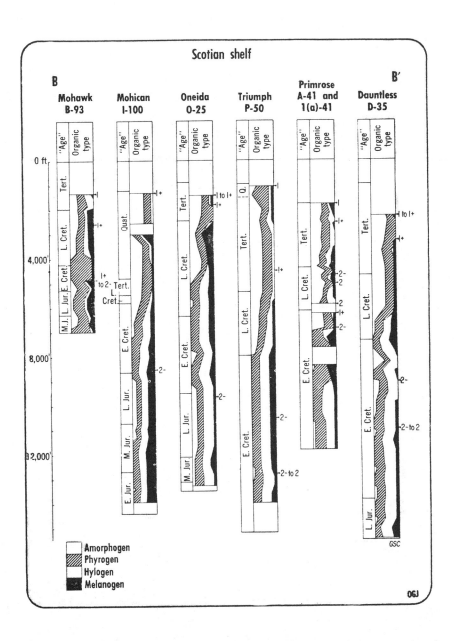

Figure 7.19: Profiles of organic matter type on a passive
margin (Bujak et al., 1977).

oils have high pour points of 79 to 85°C whereas the shallow oil (Garoupa) has a much lower pour point. On the east coast of North America organic matter-type profiles (Figure 7.19) clearly show the sequence of increasing terrestrial contribution with depth. These sediments will generate gas, or possibly some waxy crude and condensate, when buried deeply enough, but most of this area is immature. The exception seems to be in the area of Sable Island where higher geothermal gradients have led to the formation of some gas deep with oil shows shallow.

REFERENCES

Barker, C., 1974. Pyrolysis techniques for source-rock evaluation: Bull. AAPG, v. 50, p. 2349-2361.

Barker, C., 1977. Aqueous solubility of petroleum as applied to its origin and primary migration - Discussion: Bull. AAPG, v. 61, p. 2146-2149.

Bailey, N. J. L., Evans, C. R. and Milner, C. W. D., 1974. Applying petroleum geochemistry to search for oil: examples from Western Canada basin: Bull. AAPG, v. 58, p. 2284-2294.

Bray, E. E. and Evans, E. D., 1961. Distribution of n-paraffins as a clue to recognition of source beds: Geochim. Cosmochim. Acta, v. 22, p. 2-15.

Bujak, J. P., Barss, M. S. and Williams, G. L., 1977. Offshore East Canada's organic type and color and hydrocarbon potential: Oil & Gas J., April 1977, p. 198-202.

Campos, C. W. M., Muira, K. and Reis, L. A. N., 1975. The East Brazilian continental margin and petroleum prospects: Proc. 9th World Petr. Congr., v. 2, p. 71-81.

Claypool, G. E. and Reed, P. R., 1976. Thermal analysis technique for source-rock evaluation - Quantitative estimate of organic richness and effects of lithologic variation: Bull. AAPG, v. 60, p. 608-611.

Claypool, G. E., Lubeck, C. M., Baysinger, J. P. and Ging, T. G., 1977: Organic Geochemistry. p. 46-59 in "Geological Studies on the COST No B-2 well, U.S. Mid-Atlantic Outer Continental Shelf Area: (Ed., P. A. Scholle) U.S.G.S. Circ #750.

Cooper, B. S., Coleman, S. H., Barnard, P. C. and Butterworth, J. S., 1975. Paleotemperatures in the northern North Sea basin: in "Petroleum and the Continental Shelf of North-West Europe" (Ed. A. Woodland), Applied Science, Barking. v.1, p. 487-492.

Correia, M., 1971. Diagenesis of sporopollenin: p. 569-620 in "Sporopollenin" (Ed. by Brooks, Grant, Muir, Van Gijzel and Shaw), Academic Press, London.

Dow, W. G. and Pearson, D. B., 1975. Organic matter in Gulf Coast sediments: Preprints, 7th Ann. Offshore Tech. Conf., paper no. OTC 2343.

Dow, W. G., 1977. Kerogen studies and geological interpretations: J. Geochem. Expl., v. 7, p. 79-99.

Dow, W. G., 1977. Petroleum source beds on continental slopes and rises: Bull. AAPG, v. 62, p. 1584-1606.

Espitalie, J., LaPorte, J. L., Madec, M., Marquis, F., LePlat, P., Poulet, J. and Boutefeu, A., 1977. Methode rapide de characterisation des roches meres de leur potential et de leur degre d'evolution: Rev. Inst. Francais Petrol., v. 32, p. 23-42.

Evans, C. R. and Staplin, F. L., 1971. Regional facies of organic metamorphism: in R. W. Boyle and J. I. McGerrigle (Eds.) "Geochemical Exploration 1970" Can. Inst. Min. Metall., Spec. Vol., v. 11, p. 517-520.

Grayson, J. F., 1975. Relationship of palynomorph translucency to carbon and hydrocarbons in clastic sediments: Colloques International Petrographie de la Matiere Organique des Sediments: Paris (Centre National de la Recherche Scientifique), p. 261-273.

Halbouty, M. T., King, R. E., Klemme, H. D., Dott, R. H., Sr., and Meyerhoff, A. A., 1970. Factors affecting formation of giant oil and gas fields, and basin classification: "Geology of giant petroleum fields" (Ed. by M. T. Halbouty) AAPG Memoir 14, p. 528-551.

Hedberg, H. D., 1968. Significance of high wax oils with respect to genesis of petroleum: Bull. AAPG, v. 52, p. 736-750.

Ho, T. T. Y., 1977. Geological and geochemical factors controlling electron spin resonance signals in kerogen: paper presented at ASCOPE/CCOP Seminar on Generation and Maturation of Hydrocarbons in Sedimentary Basins, Manila.

Johns, W. D. and Shimoyama, A., 1972. Clay minerals and petroleum-forming reactions during burial and diagenesis: Bull. AAPG, v. 56, p. 2160-2167.

LaPlante, R. E., 1974. Hydrocarbon generation in Gulf Coast Tertiary
    sediments: Bull. AAPG, v. 58, p. 1281-1289.

Lutz, M., Laasschieter, J. P. H. and van Wijhe, D. H., 1975. Geo-
    logical factors controlling Rotliegend gas accumulations in the
    Mid-European basin: Proc. 9th World Petr. Congr., v. 2, p. 93-
    103.

Magnier, P., Oki, T. and Kartaadiputra, L. W., 1975. The Mahakam
    delta, Kalimantan, Indonesia: Proc. 9th World Petr. Congr., v. 2,
    p. 239-250.

Michelson, J. E., 1976. Miocene deltaic oil habitat, Trinidad:
    Bull, AAPG, v. 60, p. 1502-1519.

Nwachukwu, S. O., 1976. Approximate geothermal gradients in Niger
    delta sedimentary basin: Bull. AAPG, v. 60, p. 1073-1077.

Philippi, G. T., 1965. On the depth, time and mechanism of petro-
    leum generation: Geochim. Cosmochim. Acta, v. 29, p. 1021-1049.

Pusey, W. C., 1973. How to evaluate potential gas and oil source
    rocks: World Oil, April 1973, p. 71-75.

Ronov, A. B., 1958. Organic carbon in sedimentary rocks (in relation
    to the presence of petroleum): Geochemistry (translation), no. 5,
    p. 510-536.

Scholle, P. A. (Editor), 1977. Geological studies on the COST No.
    B-2 well, U.S. Mid-Atlantic outer continental shelf area: U.S.G.S.,
    Circular #750, 71 p.

Snowdon, L. R. and Roy, K. J., 1975. Regional organic metamorphism
    in the Mesozoic strata of the Sverdrup basin: Bull. Canadian
    Petrol. Geol., v. 23, p. 131-148.

Staplin, F. L., 1969. Sedimentary organic matter, organic metamor-
    phism and oil and gas occurrence: Bull. Canadian Petrol. Geol.,
    v. 17, p. 47-66.

Sutton, C., 1977. Depositional environments and their relation to
    chemical composition of Java Sea crude oils: paper presented at
    ASCOPE/CCOP Seminar in Generation and Maturation of Hydrocarbons
    in Sedimentary Basins, Manila.

Thompson, T. L., 1976. Plate tectonics in oil and gas exploration
    of continental margins: Bull. AAPG, v. 60, p. 1463-1501.

Trask, P. D. and Patnode, H. W., 1942. Source beds of petroleum:
    Amer. Assoc. Petrol. Geol., Tulsa, 566 p.

TOPIC 8

GEOCHEMICAL CORRELATION

INTRODUCTION

Crude oil is generated by the action of heat on the organic matter in source rocks and subsequently migrates to a reservoir. Because migration is such an inefficient process most of the generated bitumens remain in the source rock and this suggests that the extractable organic matter left in the rock and the crude oil in the reservoir may show similarities in their chemical compositions. Also, crude oils generated by the same source rock but reservoired in different traps should show similarities. On the other hand, crudes which were derived from different source rocks will have distinct compositions. The techniques used to relate crude oils to each other and to the source rocks which produced them consitute geochemical correlation.

Correlation is important in exploration. When an oil is correlated to its source rock the migration path is established and structures along it become important exploration targets. In the early stages of basin evaluation the presence of two unrelated crudes would indicate the presence of two source rocks. Any area with multiple source rocks is attractive. Similarly, if an oil and a source rock cannot be correlated, there must be another source rock for the crude oil and the first source may have produced other oils. Such a basin would be more attractive than one with only a single source rock.

Techniques for correlating normal crude oils are well established and are used routinely, but when one (or more) of the oils is degraded correlation becomes more difficult and special procedures may be needed.

Relating an oil to its source rock poses some special problems but correlation is frequently successful. The most difficult task is to relate a surface seep to reservoired crude at depth. Correlation techniques will be discussed in this sequence of increasing difficulty.

OIL-TO-OIL CORRELATION

A wide variety of techniques has been used in correlating crude oils, and they all try to show that the oils are sufficiently alike to be regarded as genetically related. Physical properties, including ^{O}API, color, pour point, etc., were widely used in early studies, but are not very reliable and have been replaced by detailed chemical data. Single parameters such as percent sulfur or nitrogen, carbon isotope

The API Gravities and Sulfur Contents of 549 Wyoming Crude Oils

	API range	API average	Sulfur Content - % range	Sulfur Content - % average
224 Cretaceous Oils	16 to 59	37.2	0.1 to 1.9	0.20
38 Jurassic Oils	14 to 63	34.0	0.1 to 3.9	1.10
6 Triassic Oils	18 to 29	22.3	1.7 to 3.1	2.70
281 Paleozoic Oils	11 to 62	26.2	0.1 to 4.7	2.39

Vanadium and Nickle Contents of Wind River Basin Crude Oils

Age	Formation	Field	Vanadium (V) ppm	Nickel (Ni) ppm	V/Ni
Cretaceous	Mesa Verde	Beaver Creek	0.9	3.0	0.3
	"	W. Poison Spider	0.1	0.1	1.0
	Cody	Pilot Butte	0.5	2.1	0.2
	Cody (?)	Alkali Butte*	4.0	15.0	0.3
	Muddy	Grieve	0.3	0.2	1.5
	"	Pilot Butte	0.3	0.2	1.5
	"	Poison Spider	0.8	5.1	0.2
	"	Rattlesnake Hills*	3.0	5.5	0.5
Jurassic	Nugget	Steamboat Butte	87	20	4.4
Triassic	Chugwater	Sheldon Dome	3.2	0.9	3.6
	"	N.W. Sheldon	50	8.0	6.2
	"	Poison Spider	136	30	4.5
	"	Steamboat Butte	32	7.7	4.2
	"	Dallas Dome*	108	23	4.7
Permian	Phosphoria	Circle Ridge	88	20	4.4
	"	Dallas Dome	90	19	4.7
	"	Lander	98	18	5.4
	"	Pilot Butte	54	11	4.9
	"	S. Sand Draw	38	12	3.2
	"	Winkleman Dome	94	20	4.7
Pennsylvanian	Tensleep	Big Sand Draw	16	3.8	4.2
	"	S. Casper Creek	161	32	5.0
	"	Circle Ridge	91	20	4.6
	"	Clark Ranch	167	39	4.3
	"	Dallas Dome	108	25	4.3
	"	Lander-Hudson	95	19	5.0
	"	Notches	56	9	6.2
	"	Pilot Butte	94	18	5.2
	"	Steamboat Butte	54	9	6.0
	"	Winkleman Dome	111	23	4.8
	Amsden	Circle Ridge	91	20	4.6
Mississippian	Madison	Circle Ridge	93	20	4.7

*Samples from Abandoned Casings.

Table 8.1: API gravities and sulfur, vanadium and nickel contents of Wind River basin crude oils (McIver, 1962).

ratios or vanadium-nickel ratios may be useful sometimes, but frequently
give overlapping ranges and are of limited applicability. McIver (1962)
found that sulfur content and API gravities gave overlapping ranges for
Cretaceous, Jurassic, Triassic and Paleozoic oils of the Wind River
basin, but that the Cretaceous oils all have vanadium-nickel ratios less
than 1.5, whereas the other oils had ratios of 3.2 or greater (Table
8.1). This shows only that there are at least two families of oils,
because the Jurassic and older oils may represent several families of
crude oils which happen to have similar vanadium-nickel ratios.

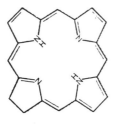

ISOPRENOID

STERANE

PORPHYRIN

Figure 8.1: Some compounds with
complex and characteristic
structures found in crude oils and
source rock extracts. Porphyrins
and steranes can have side chains
of various lengths.

The compound types used for correlation should be chosen for their
insensitivity to change during migration, maturation and alteration.
They should show sufficient complexity and variation in relative amounts
so that unrelated rocks and crude oils are unlikely to have similar
values. Some members of homologous series which fulfill these require-

ments and are used in correlation are shown in Figure 8.1. Koons et al.
(1974) used the relative amounts of various steranes to show that there
were two families of crude oils in the Tuscaloosa of Mississippi, while
porphyrins have been used successfully for oil correlation by Gransch
and Eisma (1966) in Venezuela and by Van Eggelpoel (1964) in Gabon.
Long chain isoprenoid compounds are easy to analyze and are being widely
used in correlation. The C_{19} and C_{20} chains, pristane (pr) and phytane
(ph) respectively, are quite abundant and their relative amounts show a
wide range. Powell and McKirdy (1975) found that for Australian and New
Guinea crude oils the pristane/phytane ratio ranged from 1.0 to 10.0.
The ratio appears to reflect the environment of deposition of the source
rock although ratios such as pr/nC_{17} and ph/nC_{18} sometimes give a better
indication (Lijmback, 1975; Welte et al., 1975).

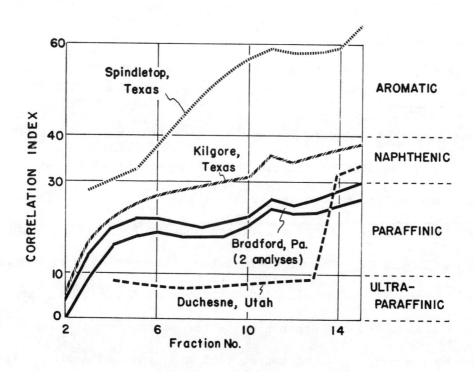

Figure 8.2: Correlation index curves of selected oils
(Barbat, 1967).

141

It is unwise to rely on a single parameter to demonstrate a relationship between oils and the most convincing correlation studies are those which have used many different types of chemical data. In 1940 Smith introduced the "Correlation Index, (C. I.)" as a way of summarizing the chemical composition of a crude oil in a single number. The index was defined in terms of the relative amount of paraffins and aromatics so that pure paraffins have a C. I. of zero and pure aromatics a C. I. of 100 with naphthenes being intermediate. The C. I. was determined for each of the 14 fractions produced from a Hempel distillation of the crude oil, and then plotted against boiling point fraction to give a characteristic curve. Similarly shaped curves were interpreted as showing that the oils are related (Figure 8.2). A large number of Correlation Indexes and correlation diagrams have been used in exploration by Barbat (1967), Stone (1967) and others, and in a modified form by Jones and Smith (1965).

With the introduction of improved analytical techniques, particularly gas chromatography, various petroleum fractions can now be analyzed for a large number of individual compounds and their relative abundances used to characterize oils. The most frequently studied material contains compounds whose boiling points are higher than the C_{15} straight-chain paraffins ("C_{15+} fraction"). The relative abundance of straight-chain paraffins within the C_{15+} fraction are often characteristic of the oil. Powell and McKirdy (1975) examined a wide range of Australian oils and found that the amount and distribution of the normal paraffins was controlled by the nature of the source rock and provided a clear distinction between oils sourced in various environments. Oils sourced by sediments deposited nearshore frequently contained sufficiently high concentrations of normal paraffins to produce high pour points ("waxy oils") - an observation consistent with previous data accumulated by Hedberg (1968). Koons et al. (1974) used the C_{15+} fraction in their study of Tuscaloosa oils mentioned above and found that the groupings were the same as those deduced from steranes and carbon isotope data.

The saturate fraction that is separated from oils for normal paraffin analysis also contains isoprenoids and these are being more frequently employed in correlation. Pristane/phytane ratios show a range of roughly 1 to 10 and are widely used (Powell and McKirdy, 1975; Welte et al., 1975; Sutton, 1977; Illich et al., 1977) while other isoprenoid ratios are becoming more important.

Other correlation studies have used various details of the chemical makeup of crude oils including ratios of the amount of saturate to aromatic compounds, trace metals (particularly nickel and vanadium), odd-even carbon chain length in the straight-chain paraffins, and infrared spectra. Isotopic data (mainly for carbon), elemental composition (%S, %N) and physical parameters (^{O}API, optical rotation) have also been used. Applications of these and other correlation techniques are sum-marized in Table. 8.2.

An excellent geochemical correlation study of crude oils in the Williston basin has been published by Williams (1974). This basin is ideally suited for developing correlation procedures because organic carbon is concentrated in a limited number of widely separated Paleozoic shales. These shales are isolated by evaporites which prevent mixing of migrating oils except locally beyond the evaporite limits (Dow, 1974). 184 oils were analyzed and 19 were shown to be bacterially degraded, 23

Figure 8.2: Summary of techniques used in geochemical correlation (Barker, 1975).

Technique	Type of correlation	Reference*
Elemental analysis		
Sulfur, %...................	Oil to oil	3 (Wyoming, Utah, Colorado); 4 (Colorado, Utah); 10 (Italy); 18 (Canada;) 21 (Texas); 33 (Wyoming); 36 (Algeria); 39 (Wyoming); 42 (Wyoming); 50 (Montana, N. Dakota)
	Oil to source	8 (U.S.)
Nitrogen, %................	Oil to oil	3 (Wyoming, Utah, Colorado); 21 (Texas, N. Mexico); 39 (Wyoming)
	Oil to source	8 (U.S.)
Nickel, vanadium.............	Oil to oil	1 (Iraq); 10 (Italy); 33 (Wyoming); 39 (Wyoming)
Trace metals (other than Ni, V).	Oil to oil	10 (Italy;) 11 (California); 18 (Canada)
	Oil to source	8 (U.S.)
Isotopic ratios		
Carbon (SC¹³)...............	Oil to oil	5 (Russian); 15 (Utah, Mississippi, North Sea); 16 (Russia); 22 (Mississippi); 24 (Texas, N. Mexico); 28 (Poland); 31 (Russia); 39 (Wyoming); 42 (Wyoming); 50 (Montana, N. Dakota)
Sulfur (SC³⁴)...............	Oil to oil	5 (Russia); 40 (Iraq); 42 (Wyoming)
Gross composition		
Saturates, %...............	Oil to oil	21 (Texas); 23 (Poland); 26 (Poland) 27 (Poland); 28 (Poland); 44 (Rumania); 45 (Germany)
	Oil to source	2 (Kansas, Oklahoma); 44 (Rumania)
Naphthenes, %.............	Oil to oil	21 (Texas, N. Mexico); 44 (Rumania)
Aromatics, %.............	Oil to oil	5 (Russia); 20 (Wyoming); 21 (Texas, N. Mexico); 23 (Poland); 26 (Poland); 28 (Poland); 39 (Wyoming)
	Oil to source	2 (Kansas, Oklahoma); 5 (Russia); 20 (Wyoming)
Detailed composition		
Saturates...................	Oil to oil	5 (Russia); 6 (Russia); 11 (California); 13 (Chile); 15 (Utah, Mississippi, North Sea); 24 (Russia); 29 (Australia); 32 (Australia); 35 (California); 36 (Algeria); 45 (Germany); 50 (Montana, N. Dakota)
	Oil to source	5 (Russia); 7 (U.S.); 8 (U.S.); 15 (Utah, Mississippi, North Sea); 30 (Australia); 31 (Russia); 35 (California); 47 (Germany); 50 (Montana, N. Dakota)
Naphthenes.................	Oil to oil	12 (Canada); 24 (Russia); 29 (Australia); 35 (California)
	Oil to source	31 (Russia); 35 (California)
Aromatics...................	Oil to oil	6 (Russia); 9 (Russia); 11 (California); 27 (Poland); 36 (Algeria); 45 (Germany)
	Oil to source	8 (U.S.); 31 (Russia)
Specific compounds		
Isoprenoids..................	Oil to oil	45 (Germany); 47 (Germany)
	Oil to source	31 (Russia)
Porphyrins..................	Oil to oil	17 (Venezuela); 41 (Gabon)
	Oil to source	17 (Venezuela)
Steranes.....................	Oil to oil	22 (Mississippi)
Miscellaneous		
Correlation Index..............	Oil to oil	3 (Wyoming, Utah, Colorado); 4 (Colorado, Utah); 21 (Texas, New Mexico); 26 (Poland); 34 (Oklahoma, Kansas); 39 (Wyoming); 50 (Montana, N. Dakota)
Optical rotation...............	Oil to oil	50 (Montana, N. Dakota)
	Oil to source	8 (U.S.)
Spectroscopic methods.........	Oil to oil	10 (Italy)
(i.r., fluorescence, etc.)	Oil to source	8 (U.S.); 19 (Utah); 37 (California, Colorado, Wyoming, Canada); 49 (Germany)
Ratios of selected compounds...	Oil to oil	15 (Mississippi, Utah, North Sea); 22 (Mississippi)
	Oil to source	15 (Mississippi, Utah, North Sea)

144

[1] Al-Shahristani, H. and Al-Atyia, M. J., "Vertical migration of oil in Iraqi oil fields: evidence based on vanadium and nickel concentrations," *Geochim. Cosmochim. Acta*, 36, 929-938, 1972.

[2] Baker, D. R., "Organic geochemistry of Cherokee group in Southeastern Kansas and Northeastern Oklahoma," *Bull. AAPG*, 46, 1621-1642, 1962.

[3] Barbat, W. N., "Crude-oil correlations and their role in exploration," *Bull. AAPG*, 51, 1255-1292, 1967.

[4] Bass, N. W., "Composition of crude oils in Northwestern Colorado and Northeastern Utah suggests local sources," *Bull. AAPG*, 47, 2039-2064, 1963.

[5] Botneva, T. A. and Maksimov, S. P., "Criteria of comparison of oils and organic matter of source rocks," (in Russian) *Geol. Nefti Gaza*, 5, 33-37, 1969.

[6] Botneva, T. A., "Types of oils of the West Fore Corpathians and their relation to conditions of oil and gas generation and occurrence," U.S.S.R. All Union Oil and Gas Genesis Symp. (Moscow, February 1967) Proc. pp. 177-178 (in Russian).

[7] Bray, E. E. and Evans, E. D., "Hydrocarbons in non-reservoir-rock source beds," *Bull. AAPG*, 49, 248-257, 1965.

[8] Brenneman, M. C. and Smith, P. V., "The chemical relationship between crude oils and their source rocks" in "Habitat of Oil," AAPG, Tulsa, Okla., pp. 818-849, 1958.

[9] Chakhmakhchev, V. A. and Vinogradova, T. L., "Geochemical peculiarities of the hydrocarbon composition of benzines of Tersk-Sunzhinsk zone crudes," *Geokhimiya*, 7, 844-850, (in Russian), 1972.

[10] Colombo, U. and Sironi, G., "Geochemical analysis of Italian oils and asphalts," *Geochim Cosmochim. Acta*, 25, 25-51, 1961.

[11] Connor, J. J. and Gerrild, P. M., "Geochemical differentiation of crude oils from six Pliocene sandstone units, Elk Hills U.S. Naval Petroleum Reserve No. 1, California," *Bull. AAPG*, 55, p. 1802-1813, 1971.

[12] Deroo, G., Tissot, B., McCrossan, R. G., and Der, F., "Geochemistry of the heavy oils of Alberta," Oil Sands Fuel of the Future. Mem. 3, Canad. Soc. Petr. Geol., 148-167, 1975.

[13] Didyk, B. M., Ober, G. and McCarthy, E. D., "Fingerprinting the migration of a Chilean oil," *Geochim. Cosmochim. Acta*, 37, 1402-1406, 1973.

[14] Dow, W. G., Application of oil-correlation and source-rock data to exploration in Williston basin," *Bull. AAPG*, 58, 1253-1262, 1974.

[15] Erdman, J. G. and Morris, D. A., "Geochemical correlation of petroleum," *Bull. AAPG*, 58, 2326-2337, 1974.

[16] Galimov, E. M., Kuznetsova, N. G., Pyankov, N. A. and Vinnikovskii, S. A., "Genetic types of Perm Prekama crudes on the basis of carbon isotope composition," *Geol. Nefti Gaza.*, 1, 33-39 (in Russian), 1972.

[17] Gransch, J. A. and Eisma, E., "Geochemical aspects of the occurrence of porphyrins in West Venezuelan mineral oils and rocks," *Advances in Organic Geochemistry*, Ed. by G. D. Hobson and G. C. Speers, 69-86, 1966.

[18] Hobson, G. W., "Vanadium, nickel and iron trace metals in crude oils of Western Canada," *Bull. AAPG*, 38, 1671-1698, 1954.

[19] Hunt, J. M., Steward, F. and Dickey, P., "Origin of hydrocarbons of Uinta basin, Utah," *Bull. AAPG*, 38, 1671-1698, 1954.

[20] Hunt, J. M. and Jamieson, G. W., "Oil and organic matter in source rocks of petroleum," *Bull. AAPG*, 40, 477-488, 1956.

[21] Jones, T. S. and Smith, H. M., "Relationships of oil composition and stratigraphy in the Permian basin of West Texas and New Mexico," *Fluids in Subsurface Environments* Memoir #4 AAPG, 101-224, 1965.

[22] Koons, C. B., Bond, J. G. and Peirce, F. L., "Effects of depositional environment and past depositional history on chemical composition of Lower Tuscaloosa oils," *Bull. AAPG*, 58, 1272-1280, 1974.

[23] Kozikowski, H. and Marzec, A., "A new method of geochemical correlation of crude oils for geological defining of the optimum exploration zones in the Carpathians and the Carpathian foreland," (in Polish) *Nafta (Pol.)*, 27, 134-147, 1971.

[24] Kvenvolden, K. A. and Squires, R. M., "Carbon isotopic composition of crude oils from Ellenburger Group (Lower Ordovician), Permian basin, West Texas and Eastern New Mexico," *Bull. AAPG*, 51, 1293-1303, 1967.

[25] Maksimov, S. P. and Sofonova, G. I., "Genetic characteristics of petroleums in Devonian deposits of the Volga-Urals province," *Int. Geol. Rev.*, 15, 497-507, 1973.

[26] Marzec, A., Kisielow, W. and Orzechowski, P., "Comparison of the results of the crude oil correlation based on Y — F(x) function and on the correlation index curves," *Advances in Organic Geochemistry*, Ed. by Pergamon Press, 517-522, 1972.

[27] Marzec, A. and Burszyk, R., "Correlation of the crude oils based on the definite variability of chemical composition of the groups of crudes," *Advances in Organic Geochemistry*, Ed. by Tissot and Bienner, 1973.

[28] Marzec, A., Kozibowski, H., Glogoczowski, J. and Kisielow, W., "Problems of migration of Polish crude oils on the basis of geochemical correlation data and carbon isotope composition," *Chem. Geol.*, 8, 197-217.

[29] Mathews, R. T., Burns, B. J. and Johns, R. B., "Comparison of hydrocarbon distribution in crude oils and shales from Moonie field, Queensland, Australia," *Bull. AAPG*, 54, 428-438, 1970.

[30] Mathews, R. T., Burns, B. J. and Johns, R. B., "An approach to identification of source rocks," *J. Australian Petrol. Explor. Assoc.*, 11, 115-120, 1971.

[31] Maximov, S. P., Botneva, T. A., Rodionova, K. Ph., Larskaya, E. S. and Sofonova, G. I., "Genetic criteria for comparison of oil with organic matter," *Advances in Organic Geochemistry*, Ed. by Tissot and Bienner, 1973.

[32] McKirdy, D. M. and Powell, T. G., "Crude oil correlations in the Perth and Carnarvon basins," *J. Australian Petrol. Explor. Assoc.*, 13, 81-85, 1973.

[33] McIver, R. D., "The crude oils of Wyoming—product of depositional environment and alteration," in Symp. on Early Cret. Rocks of Wyoming and Adjacent Areas: Wyoming Geol. Assoc. 17th Annual Field Conference, 248-251, 1962.

[34] Neumann, L. M., et al., "Relationship of crude oils and stratigraphy in parts of Oklahoma and Kansas," *Bull. AAPG*, 31, 92-148, 1947.

[35] Philippi, G. T., "On the depth, time, and mechanism of petroleum generation," *Geochim. Cosmochim. Acta*, 22, 1021-1040, 1965.

[36] Poulet, M. and Roucache, J., "Geochemical study of the north Sahara reservoirs of Algeria," *Rev. Inst. Francais Petr.*, May 1969, 616-664, 1969.

[37] Riecker, R. E., "Hydrocarbon fluorescence and migration of petroleum," *Bull. AAPG*, 46, 60-75, 1962.

[38] Smith, H. M., "Correlation Index to aid in interpreting crude oil analyses," U.S. Bur. Mines Tech. Paper 610, 34 p., 1940.

[39] Stone, D. S., "Theory of palezoic oil and gas accumulation in Big Horn basin, Wyoming," *Bull. AAPG*, 51, 2056-2114, 1967.

[40] Thode, H. G. and Monster, J., "Sulfur isotope abundances and genetic relations of oil accumulations in Middle East basin," *Bull. AAPG*, 54, 627-637, 1970.

[41] van Eggelpoel, A., "Comparison entre les porphyrins extraites de la roche reservoir (argiles silicifiees) d'Ozouri (Gabon) et celles de l'huile," Paper presented at 2nd Intl. Geochem. Congr., Paris, 1964.

[42] Vredenburgh, L. D. and Cheney, E. S., "Sulfur and carbon isotopic investigations of petroleum, Wind River basin, Wyoming," *Bull. AAPG*, 55, 1954-1975, 1971.

[43] Wagner, F. J. and Iglehart, C. F., "North American drilling activity in 1973," *Bull. AAPG*, 58, 1475-1505, 1974.

[44] Walters, R. P., "Pliocence oils of Rumania," *Bull. AAPG*, 53, 2190-2194, 1969.

[45] Wehner, H., "Crude oil chemistry and its relation to oil migration history of the lower Saxony basin (Federal Republic of Germany)," Ed. by Tissot and Bienner, 409-421, 1973.

[46] Welte, D. H., "Relation between petroleum and source rock," *Bull. AAPG*, 49, 2246-2268, 1965.

[47] Welte, D. H., "Correlation problems among crude oils," *Advances in Organic Geochemistry*, Ed. by G. D. Hobson and G. C. Speers, 111-127, 1966.

[48] Welte, D. H., "Petroleum exploration and organic geochemistry," *J. Geochem. Explor.*, 1, 117-136, 1972.

[49] Wieseneder, H., "The problem of source rocks for oil in the Vienna basin," *Erdoel Z*, 80, 479-486 (in German), 1968.

[50] Williams, J. A., "Application of oil-correlation and source-rock data to exploration in the Williston basin," *Bull. AAPG*, 58, 1243-1252, 1974. ∎

were mixtures of two other types and 17 oils were thought to be produced from minor local sources and to be unrelated to the major types. The remaining 125 oils were divided into three groups (Types I-III) based on a variety of geochemical information. Sulfur percentages and API gravities for the oils produced no distinct groupings but carbon isotope values showed that there were at least two families of oils in the basin. Correlation Index curves suggested three different families and this was supported by the nature of normal-paraffin abundances as a function of chain length (Figure 8.3). These groupings coincided with those obtained by plotting C_4-C_7 gas chromatographic data on a branched chain-straight chain-naphthene ternary diagram. 42 oils were assigned to Type I and these were reservoired mainly in Ordovician age rocks. Type II had 76 oils reservoired predominantly in Mississippian age reservoirs while the 7 oils assigned to Type III all occurred in Pennsylvanian age rocks. This geochemical information was integrated into the geologic framework by Dow (1974) who concluded that "the addition of geochemical information to exploration programs can contribute significantly to a better under-standing of the habitat of oil and the factors which control its distri-bution. This can result in improved exploration success ratios, even in relatively highly explored geologic provinces such as the Williston basin."

CORRELATION OF ALTERED OILS

Correlation can be difficult when the composition of one (or more) of the oils has been changed by maturation or alteration. Thermal maturation and biodegradation produce well-documented trends relative to the unaltered oil (Table 8.3) and show that parameters such as percent paraffins cannot be used. Biodegraded oils must be correlated using

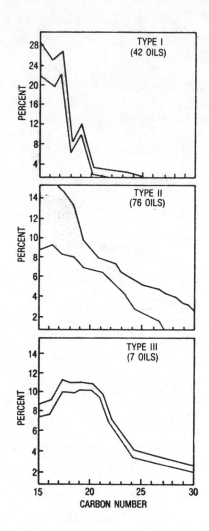

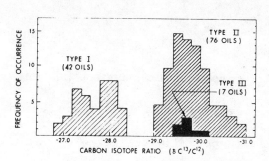

Carbon-isotope ratios for three basic types of Williston basin oils. Commingled oils and microbially altered oils are not included.

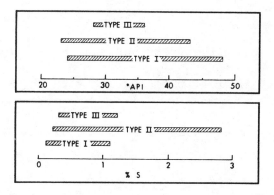

Ranges of API gravity and percent sulfur for each of three basic oil types in Williston basin.

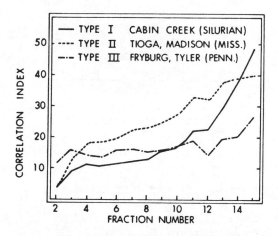

Typical correlation-index curves for unaltered Williston basin oils (from Bureau of Mines data). Each curve represents one oil sample of designated type.

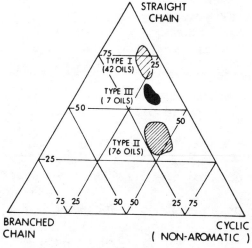

Hydrocarbon-type distribution in C_4-C_7 fraction, Williston basin oils.

Figure 8.3:   Geochemical data for the major oil types in the Williston basin (Williams, 1974).

compounds that resist bacterial breakdown. In the early stages of degradation useful correlation parameters include isoprenoid ratios (especially pristane/phytane), δC^{13} values and nickel-vanadium ratios, but for more severely degraded crudes multiringed naphthenes, sulfur compounds or higher aromatics must be used. The shapes of correlation index plots are changed by alteration and maturation but the changes occur in predictable ways. Water washing preferentially removes the aromatics from the light ends and depresses the curve for the low boiling fractions, while bacterial alteration has the opposite effect because the paraffins are removed (Figure. 8.4).

Table 8.3: Comparison of thermal and microbial alteration of crude oil.

Property	Thermal alteration	Microbial alteration
Paraffins	Increase	Decrease
Organic sulfur	Decrease	Increase
Organic nitrogen	Decrease	Increase
Light aromatics	Increase	Decrease
API gravity	Increase	Decrease
Asphaltenes	Decrease	Increase
Optical activity	Decrease	Increase

Crude oils reservoired in the Cretaceous of the Maranon basin, Peru show various levels of n-paraffin loss, and Illich et al. (1977) have divided the oils into two types on this basis. Type I oils contain high percentages of n-paraffins and occur in the western part of the basin while the Type 2 oils have much reduced concentrations of n-paraffins and occur on the eastern flanks of the basin (Figure 8.5). The relative abundances of the isoprenoids were used to show that both oil types were related. They concluded that heavier (and altered) oils are more likely to be found in the eastern part of the basin as exploration drilling proceeds.

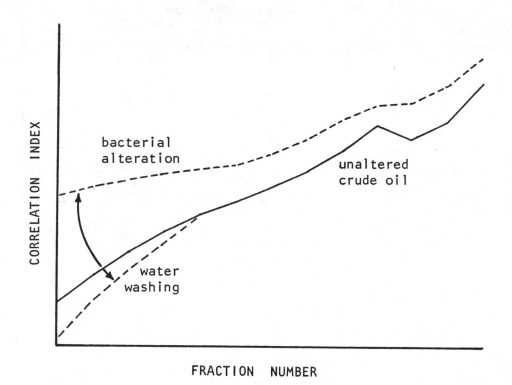

Figure 8.4: Schematic representation of the effects of water washing and bacterial alteration on Correlation Index plots.

Bacteria remove some of the single ringed naphthenes (cyclopentanes and cyclohexanes) along with the normal paraffins (Figure 8.6) and these compound types cannot be used in correlating degraded oils. The multi-ringed naphthenes however resist bacterial degradation and have been used in correlation. Deroo et al. (1975) reported a successful grouping of degraded Canadian oils using 3-, 4-, and 5-ring naphthene ternary plots* (Figure 8.7). It should be noted that the steranes which are widely used in correlation (e.g. Koons et al., 1974) are a particular

*The corresponding gas chromatograms are given in Figure 6.11 of Topic 6.

C R I T E R I A	OIL TYPE			
	1A	1B	2A	2B
ABUNDANT NORMAL PARAFFINS	X	X		
MINOR NORMAL PARAFFINS			X	X
METHYLCYCLOHEXANE MOST ABUNDANT	X			
N-HEPTANE MOST ABUNDANT		X		
NORMAL PARAFFINS > ISOPRENOIDS	X	X	X	
ISOPRENOIDS ≅ NORMAL PARAFFINS				X

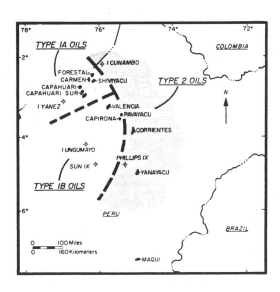

Figure 8.5: Generalized distribution of the major oil types in northeastern Peru.

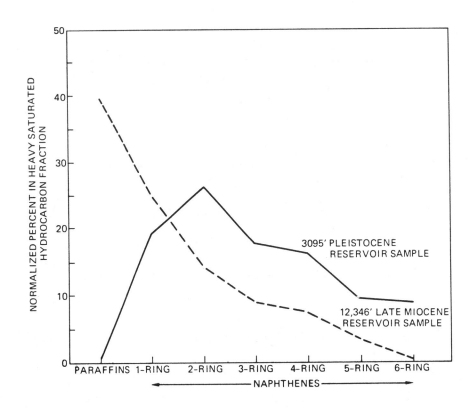

Figure 8.6: Relative amounts of paraffins and 1 to 6-ring naphthenes in C_{15} - plus fraction of two West Delta Block 30 oil samples; solid line, 3,095-ft. (943 m) Pleistocene reservoir sample; dashed line, 12,346-ft. (3,763 m) late Miocene reservoir sample (Young et al., 1977).

type of 4-ring naphthene and can be used with degraded oils.

The carbon isotope ratios for oils are not changed appreciably by biodegradation and Sutton (1977) found that in Java Sea oils the δC^{13} value could be used to correlate degraded oils to equivalent undegraded oils. He examined 26 oils from the area north and west of Madura and

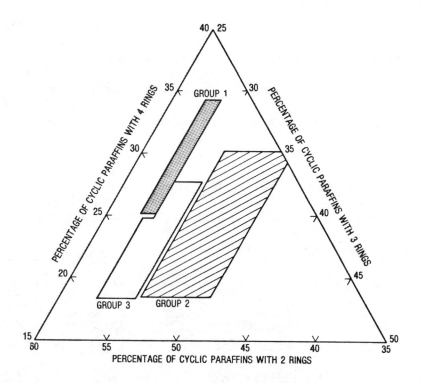

Figure 8.7: Distribution of cyclic paraffins with two, three and four rings in bacterially altered crude oils from Canada (Deroo et al., 1975). Group 1: post-Mannville oils; Group 2: Mannville oils; Group 3: Devonian oils. (See also Deroo et al., 1977).

established the presence of three types on the basis of carbon isotope values for saturate and aromatic fractions, the pristane/phytane ratios and the character of the n-paraffin distributions. The validity of these assigned groupings was confirmed by multivariate discriminant statistical analysis. δC^{13} values for total oils have overlapping ranges. Plots of isotopic ratios for saturate and aromatic factions can separate the oils into three groups. These are shown in Figures 8.8 and 8.9 along with isotopic values for the two altered oils.

SOURCE ROCK - CRUDE OIL CORRELATION

Correlation of a crude oil to the source rock which produced it is carried out by comparing the chemical composition of the oil with that of the source rock extract. Since extractable hydrocarbons are present in small amounts the experimental techniques are limited to those using small samples, and Correlation Index plots, API gravity and optical rotation are not normally applicable. Correlations based on the similarity in composition of the rock extract and oil assume that the migration process does not change the composition. A further complication arises because shale source rocks are non-uniform and may vary in composition from point to point, while crude oils are the average product from the whole shale unit. In spite of these difficulties, source rocks are being identified for an increasing number of crude oils.

In the Williston basin study discussed above, geochemical correlation procedures led to the identification of the source rocks for each of the three crude oil families. All the major shale units in the basin were analyzed and some did not have extracts resembling any of the crude oils. For example, geologic considerations do not eliminate the Heath shale as a possible source for Type III oils but the composition of the C_4-C_7 extract is not like that for the oils whereas the extract from the Tyler shale resembles the oils rather closely (Figure 8.10). The assignment of source rock to each of the three oil types was based on C_4-C_7 and C_{15+} chromatographic data and carbon isotope ratios. Sutton's study of Java Sea oils also led to the identification of source rocks.

152

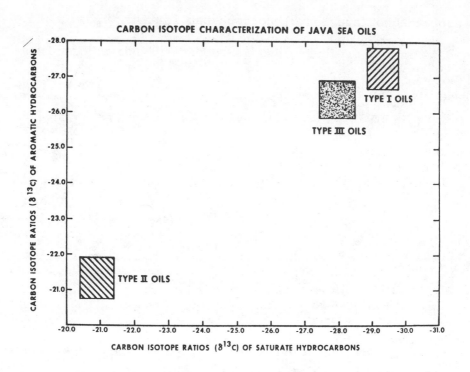

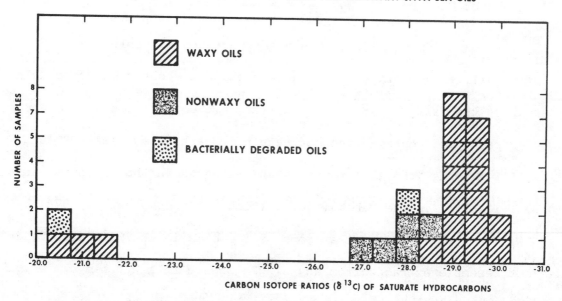

Figures 8.8 and 8.9: Carbon isotope distributions of waxy and
nonwaxy Java sea oils.

In this case C_{15+} n-paraffin data were combined with isoprenoid ratios

and carbon isotope data to make the correlations. Other examples of

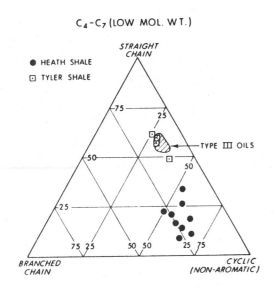

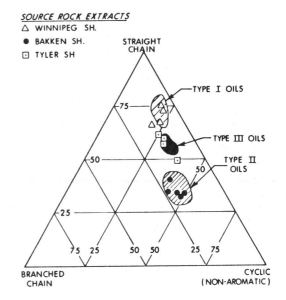

Figure 8.10:

Hydrocarbon type distribution in C_4-C_7 fraction of Heath and Tyler extracts, compared to Type III oils.

Hydrocarbon type distribution in C_4-C_7 fraction; comparison of Williston basin basic oil types with source-bed extracts.

crude oil-source rock correlations are referenced in Table 8.2.

The Williston basin study is an exceptionally clear cut case. In most areas the oils do not have quite the same composition as the source rock extracts and contain more saturate hydrocarbons and less NSO's (see, for example, the data for oils and extracts in the Parentis Basin (Deroo, 1976) or in the Mahakam delta (Combaz and deMatharel, 1978)).

Williams (1974) found that some of the oils in the Williston basin did not resemble extracts from any single source rock but had characteristics intermediate between those of two source rocks and presented evidence to show that oil in the Weldon field came from both Bakken and Winnipeg sources. A similar situation has been documented for the Egret field in Australia (Powell, 1975).

As new analytical techniques are developed they are applied in correlation. Recently gas chromatographs with sulfur specific detectors (called "Flame Photometric Detectors" (FPD)) have been developed. They give a gas chromatogram of only the sulfur compounds which provides a "fingerprint" of the oil or rock extract. Correlations are based on similarities between sulfur fingerprints. Oudin (1976) used this technique in a study of North Sea oils (along with other geochemical techniques) to show that the Jurassic oils are similar to hydrocarbon extracts from the Kimmeridgian.

SURFACE SEEP-OIL CORRELATIONS

Seeps are the surface expression of crude oil at depth and as such are important in the early stages of exploration in new areas. Unfortunately, surface seeps provide an extreme example of degraded oils with losses occurring due to evaporation, water leaching, bacterial degradation and oxidation. Any compound or group of compounds sensitive to these influences cannot be used for correlation. For example, hydrocarbons below C_{10} are lost, and normal paraffins throughout the entire range are usually missing. Percentages of sulfur and nitrogen are increased and API gravity lowered. In favorable cases correlation can be carried out based on carbon isotope ratios or nickel/vanadium ratios. Compositional information from infrared spectra and the relative amounts of multi-ringed naphthenes may also be useful for correlation in these cases.

CORRELATION OF GASES

Gases are difficult to correlate to their source rocks and to other gases because they have a rather simple composition lacking complex molecules, and because they are very mobile and appreciable compositional changes occur during secondary migration. Isotopic composition of individual hydrocarbon components have been used and when combined with chemical and isotopic data for the associated inorganic gases often

provide a reliable method for correlation (Stahl, 1977).

REFERENCES

Barbat, W. N., 1967. Crude-oil correlations and their role in exploration: Bull. AAPG, v. 51, p. 1255-1292.

Barker, C., 1975. Oil source rock correlation aids drilling site selection: World Oil, Oct. 1975, p. 121-126, 213.

Combaz, A. and deMatharel, M., 1978. Organic sedimentation and genesis of petroleum in Mahakam Delta, Borneo: Bull. AAPG, v. 62, p. 1684-1695.

Deroo, G., Tissot, B., McCrossan, R. G. and Der, F., 1975. Geochemistry of the heavy oils of Alberta: Oil Sands of the Future, Mem. #3, Canadian Soc. Petrol. Geol., p. 148-167 (see also Deroo et al., 1977).

Deroo, G., 1976. Correlations of crude oils and source rocks in some sedimentary basins (in French): Bull. Centre Rech. Pau-SNPA, v. 10, p. 317-335.

Deroo, G., Powell, T. G., Tissot, B. and McCrossan, R. G., 1977. The origin and migration of petroleum in the Western Canadian sedimentary basin, Alberta - A geochemical and thermal maturation study: Geol. Surv. Canad., Bull. #262, 136 p.

Dow, W. G., 1974. Application of oil-correlation and source-rock data to exploration in Williston basin: Bull. AAPG, v. 58, p. 1253-1262.

Gransch, J. A. and Eisma, E., 1970. Geochemical aspects of the occurrence of porphyrins in West Venezuelan mineral oils and rocks: in G. D. Hobson and G. C. Speers (Eds.) "Advances in Organic Geochemistry" Pergamon, p. 69-86.

Hedberg, H. D., 1968. Significance of high wax oils with respect to genesis of petroleum: Bull. AAPG, v. 52, p. 736-750.

Illich, H. A., Haney, F. R. and Jackson, T. J., 1977. Hydrocarbon geochemistry of oils from Maranon basin, Peru: Bull. AAPG, v. 61, p. 2103-2114.

Jones, T. W. and Smith, H. M., 1965. Relationships of oil composition and stratigraphy in the Permian basin of West Texas and New Mexico: AAPG Memoir #4, p. 101-224.

Koons, C. B., Bond, J. G. and Peirce, F. L., 1974. Effects of depositional environment and post depositional history on chemical composition of Lower Tuscaloosa oils: Bull. AAPG, v. 58, p. 1271-1280.

Lijmbach, G. W. M., 1975. On the origin of petroleum: Proc. 9th World Petrol. Conf., Applied Science Publishes, London, v. 2, p. 357-369.

McIver, R. D., 1962. The crude oils of Wyoming - product of depositional environment and alteration: <u>in</u> Symp. on Early Cret. Rocks of Wyoming and Adjacent Areas: Wyoming Geol. Assoc. 17th Ann. Field Conf., p. 248-251.

McKirdy, D. M. and Powell, T. G., 1973. Crude oil correlation in the Perth and Carnarvon basins: J. Austral. Petrol. Expl. Assoc., v. 13, p. 81-85.

Oudin, J-L., 1976. Geochmical study of the North Sea basin (in French): Bull. Centre Rech. Pau-SNPA, v. 10, p. 339-350.

Powell, T. G., 1975. Geochemical studies related to the occurrence of oil and gas in the Dampier sub-basin, Western Australia: J. Geochem. Explor., v. 4, p. 441-466.

Powell, T. G. and McKirdy, D. M., 1975. Geologic factors controlling crude oil composition in Australia and Papua, New Guinea: Bull. AAPG, v. 59, p. 1176-1197.

Smith, H. M., 1940. Correlation Index to aid in interpreting crude-oil analyses: Bu. Mines Tech. Paper 610, 34 pp.

Stahl, W. J., 1977. Carbon and nitrogen isotopes in hydrocarbon research and exploration: Chem. Geol., v. 20, p. 121-149.

Stone, D. S., 1967. Theory of paleozoic oil and gas accumulation in Big Horn basin, Wyoming: Bull. AAPG, v. 51, p. 2056-2114.

Sutton, C., 1977. Depositional environments and their relation to chemical composition of Java Sea crude oils: paper presented at ASCOPE/CCOP Seminar on Generation and Maturation of Hydrocarbons in Sedimentary Basins, Manila.

van Eggelpoel, A., 1964. Comparison entre les porphyrins extraites de la rocke reservoir (argiles silicifices) d'Ozouri (Gabon) et celles de l'huile: paper presented at 2nd Intl. Geochem. Congr., Paris.

Welte, D. H., Hagemann, H. W., Hollerbach, A. and Leythhaeuser, D., 1975. Correlation between petroleum and source rock: Proc. 9th World Petr. Congr., v. 2, p. 357-369.

Williams, J. A., 1974. Application of oil-correlation and source rock data to exploration in the Williston basin: Bull. AAPG, v. 58, p. 1243-1252.

Young, A., Monaghan, P. H. and Schweisberger, R. T., 1977. Calculation of ages of hydrocarbons in oils - physical chemistry applied to petroleum geochemistry I: Bull. AAPG, v. 61, p. 573-600.

APPENDIX

COMMON ABBREVIATIONS

a.m.u.	atomic mass units
arom	aromatics
bbl	barrel
CI	Correlation Index; chemical ionization
CPI	Carbon Preference Index
d.a.f.	dry ash free
D.O.M.	dissolved organic matter
esr	electron spin resonance
FID	flame ionization detector
FPD	flame photometric detector
GC	gas chromatography
GLC	gas-liquid chromatography
GC-MS	gas chromatograph-mass spectrometer combination
GOR	gas-oil ratio
IP	isoprenoid
i.r.	infrared
LOM	level of organic maturity
m.a.f.	mineral ash free
MS	mass spectrometer
NSOs	nitrogen-sulfur-oxygen compounds
O.C.	organic carbon
O.M.	organic matter
PAH	polynuclear aromatic hydrocarbons
PDB	Pee Dee belemnite (carbon isotope standard)
ppm	parts per million
R_o	percent reflectance (in oil)
sat	saturate
SOM	sedimentary organic matter
TAI	thermal alteration index
TCD	thermal conductivity detector
TD	total depth
TEA	thermal evolution analysis
TOC	total organic carbon
TOM	total organic matter
u.v.	ultra violet